Advances in High-Efficiency LLC Resonant Converters

Advances in High-Efficiency LLC Resonant Converters

Special Issue Editor

Jeehoon Jung

MDPI • Basel • Beijing • Wuhan • Barcelona • Belgrade • Manchester • Tokyo • Cluj • Tianjin

Special Issue Editor
Jeehoon Jung
Ulsan National Institute of
Science and Technology (UNIST)
Korea

Editorial Office
MDPI
St. Alban-Anlage 66
4052 Basel, Switzerland

This is a reprint of articles from the Special Issue published online in the open access journal *Energies* (ISSN 1996-1073) (available at: https://www.mdpi.com/journal/energies/special_issues/ LLC_Resonant_Converters).

For citation purposes, cite each article independently as indicated on the article page online and as indicated below:

LastName, A.A.; LastName, B.B.; LastName, C.C. Article Title. *Journal Name* **Year**, *Article Number*, Page Range.

ISBN 978-3-03928-386-6 (Hbk)
ISBN 978-3-03928-387-3 (PDF)

Contents

About the Special Issue Editor . . . . **vii**

Preface to "Advances in High-Efficiency LLC Resonant Converters" . . . . **ix**

HwaPyeong Park, DoKyoung Kim, SeungHo Baek and JeeHoon Jung
Extension of Zero Voltage Switching Capability for CLLC Resonant Converter
Reprinted from: *Energies* **2019**, *12*, 818, doi:10.3390/en12050818 . . . . **1**

Yu-Chen Liu, Chen Chen, Kai-De Chen, Yong-Long Syu and Meng-Chi Tsai
High-Frequency LLC Resonant Converter with GaN Devices and Integrated Magnetics
Reprinted from: *Energies* **2019**, *12*, 1781, doi:10.3390/en12091781 . . . . **15**

Yann E. Bouvier, Diego Serrano, Uroš Borović, Gonzalo Moreno, Miroslav Vasić, Jesús A. Oliver, Pedro Alou, José A. Cobos and Jorge Carmena
ZVS Auxiliary Circuit for a 10 kW Unregulated LLC Full-Bridge Operating at Resonant Frequency for Aircraft Application
Reprinted from: *Energies* **2019**, *12*, 1850, doi:10.3390/en12101850 . . . . **35**

HwaPyeong Park, Mina Kim, HakSun Kim and JeeHoon Jung
Design Methodology of Tightly Regulated Dual-Output LLC Resonant Converter Using PFM-APWM Hybrid Control Method
Reprinted from: *Energies* **2019**, *12*, 2146, doi:10.3390/en12112146 . . . . **55**

Michael Heidinger, Qihao Xia, Rainer Kling and Wolfgang Heering
Current Mode Control of a Series LC Converter Supporting Constant Current, Constant Voltage (CCCV)
Reprinted from: *Energies* **2019**, *12*, 2793, doi:10.3390/en12142793 . . . . **75**

Hiroki Watanabe, Jun-ichi Itoh, Naoki Koike and Shinichiro Nagai
PV Micro-Inverter Topology Using LLC Resonant Converter
Reprinted from: *Energies* **2019**, *12*, 3106, doi:10.3390/en12163106 . . . . **93**

Yuan-Chih Lin, Ding-Tang Chen and Ching-Jan Chen
Flux-Balance Control for LLC Resonant Converters with Center-Tapped Transformers
Reprinted from: *Energies* **2019**, *12*, 3211, doi:10.3390/en12173211 . . . . **105**

Hussain Humaira, Seung-Woo Baek, Hag-Wone Kim and Kwan-Yuhl Cho
Circuit Topology and Small Signal Modeling of Variable Duty Cycle Controlled Three-Level LLC Converter
Reprinted from: *Energies* **2019**, *12*, 3833, doi:10.3390/en12203833 . . . . **123**

Dongkwan Yoon, Sungmin Lee and Younghoon Cho
Design Considerations of Series-Connected Devices Based LLC Converter
Reprinted from: *Energies* **2020**, *13*, 264, doi:10.3390/en13010264 . . . . **145**

About the Special Issue Editor

Jeehoon Jung (Associate Professor, Ph.D.) received his B.S. degree in electronic and electrical engineering and his M.S. and Ph.D. degrees in electrical and computer engineering from the Department of Electronics and Electrical Engineering, Pohang University of Science and Technology (POSTECH), Pohang, South Korea, in 2000, 2002, and 2006, respectively. From 2006 to 2009, he was a Senior Research Engineer in the Digital Printing Division, Samsung Electronics Company Ltd., Suwon, South Korea. From 2009 to 2010, he was a Postdoctoral Research Associate in the Department of Electrical and Computer Engineering, Texas A&M University at Qatar (TAMUQ), Doha, Qatar. From 2011 to 2012, he was a Senior Researcher in the Power Conversion and Control Research Center, HVDC Research Division, Korea Electrotechnology Research Institute (KERI), Changwon, South Korea. From 2013 to 2016, he was an Assistant Professor in the School of Electrical and Computer Engineering, Ulsan National Institute of Science and Technology (UNIST), Ulsan, South Korea, where he is currently an Associate Professor. His research interests include dc-dc and ac-dc converters, switched-mode power supplies, motor diagnosis systems, digital control, and signal processing algorithms, power conversion for renewable energy, and real-time and power hardware-in-the-loop (HIL) simulations of renewable energy and power grids. Recently, he has been researching high-frequency power converters using wide bandgap devices, smart power transformers for smart grids, power control algorithms and power line communications for dc microgrids, and induction heating techniques for home appliances. Dr. Jung is a Senior Member of the IEEE Industrial Electronics Society, the IEEE Power Electronics Society, the IEEE Industry Applications Society, and the IEEE Power and Energy Society. He served as a Member of the Editorial Committee of the Korea Institute of Power Electronics (KIPE), Associate Editor of the Journal of Power Electronics (JPE), and now he serves as a Member of the Board of Directors in the KIPE. In addition, he is an Editorial Member of *Energies* in MDPI.

Preface to "Advances in High-Efficiency LLC Resonant Converters"

LLC resonant converters have been widely used in industrial fields because of their high efficiency, simple structure, and cost-effectiveness. Many advanced technologies and approaches have been introduced and proposed to improve its power conversion efficiency, dynamic performance, stability, reliability, etc. using enhanced devices such as wide band-gap power switches and high-speed controllers. In addition, new and advanced control algorithms have been applied to the LLC resonant converters.

This book collects several research papers related to the LLC resonant converter, which were published in a Special Issue of *Energies* on the subject area of "Advances in High-Efficiency LLC Resonant converter". This Special Issue focused on emerging power electronic topologies related to the LLC resonant converter, and its design methodology and control algorithms. Topics of interest for publication include the following:

- LLC resonant topologies;

- Resonant tank design methodology for high efficiency;

- Power loss analysis in LLC resonant converter;

- High-frequency magnetics in LLC resonant converter;

- Wide band-gap devices applied to LLC resonant converter;

- Advanced control algorithm for LLC resonant converter

I believe that all the papers in this book will inspire further research topics and other researchers who are participating in the research of the LLC resonant converter. It was my pleasure to contribute to the Special Issue of *Energies* as a Guest Editor.

Jeehoon Jung
Special Issue Editor

Article

Extension of Zero Voltage Switching Capability for CLLC Resonant Converter

HwaPyeong Park [1], DoKyoung Kim [2], SeungHo Baek [2] and JeeHoon Jung [1],*

[1]   Ulsan National Institute of Science and Technology (UNIST), Ulsan KS017, Korea; darkrla6@unist.ac.kr
[2]   LIG Nex1, Seongnam KS009, Korea; kimdokyoung@lignex1.com (D.K.); seungho.baek@lignex1.com (S.B.)
*   Correspondence: jhjung@unist.ac.kr; Tel.: +82-052-217-2140

Received: 18 January 2019; Accepted: 27 February 2019; Published: 1 March 2019

Abstract: TheCLLC resonant converter has been widely used to obtaina high power conversion efficiency with sinusoidal current waveforms and a soft switching capability. However, it has a limited voltage gain range according to the input voltage variation. The current-fed structure canbe one solution to extend the voltage gain range for the wide input voltage variation, butit has a limited zero voltage switching (ZVS) range. In this paper, the current-fed CLLC resonant converter with additional inductance is proposed to extend the ZVS range. The operational principle is analyzed to design the additional inductance for obtaining the extended ZVS range. The design methodology of the additional inductance is proposed to maximize the ZVS capability for the entire load range. The performance of the proposed method is verified with a 20 W prototype converter.

Keywords: resonant converter; bidirectional power conversion; zero voltage switching; asymmetric pulse width modulation

1. Introduction

Recently, the small sized power converterhas become significant in various industries, such as lightings, TVs, computers, and other home appliances [1,2]. A power converter operating at a high switching frequency is one effective method to improve power density [3–7]. However, the high switching frequency operation induces a large switching loss for the turn-on and turn-off states. Therefore, a soft switching capability is important to obtain a high power conversion efficiency in a high switching frequency operation [8–10]. Resonant power converters, such as LC resonance, LLC resonance, CLLC resonance, and CLL resonance, can implement the soft switching capability by the resonance, which can be a good candidate to implement the high switching frequency operation [11–15]. In addition, the wide band-gap device (WBD), such as gallium nitride (GaN) and silicon carbide (SiC), can increase the switching frequency up to several MHz compared with conventional silicon (Si)-based switching devices [16–19].

The CLLC resonant converter operating at the inductive region can obtaina zero voltage switching (ZVS) capability [20–23]. However, the large input voltage variation by the batteryinduces large switching frequency variation. In addition, the voltage gain fluctuation makes a non-ZVS operation withthe capacitive operation of the converter [24]. Therefore, the CLLC resonant converter has poor voltage gain characteristics according to the wide input voltage variation. The current-fed CLLC resonant converter can overcome the input voltage variation by the battery, since the current-fed structure compensates the input voltage variation [25,26]. However, the current-fed structure makes a limited ZVS capability of the low side switch, since the low side switch operates as the boost converter and resonant converter simultaneously. Therefore, the increase of the ZVS capability is significant to obtain a high power conversion efficiency for the entire load condition.

In this paper, the current-fed CLLC resonant converter employing the additional inductance is proposed to improve the ZVS capability. The operational principle of the proposed additional inductance is analyzed with the theoretical waveforms. From the operational principle, the design methodology of the additional inductance is analyzed to obtain the ZVS capability for the entire input voltage range and load conditions. The experimental results with a 20 W prototype converter verify the validity of the proposed additional inductance.

2. Operational Principle

Figure 1 shows the scheme of the proposed current-fed CLLC resonant converter. The current-fed structure regulates the wide input voltage variation with the asymmetric pulse width modulation (APWM), because it operates as the synchronous boost converter. The CLLC resonant tank provides the galvanic isolation using the transformer and resonance. The voltage doubler structure of the secondary side can reduce the turn ratio of the transformer.

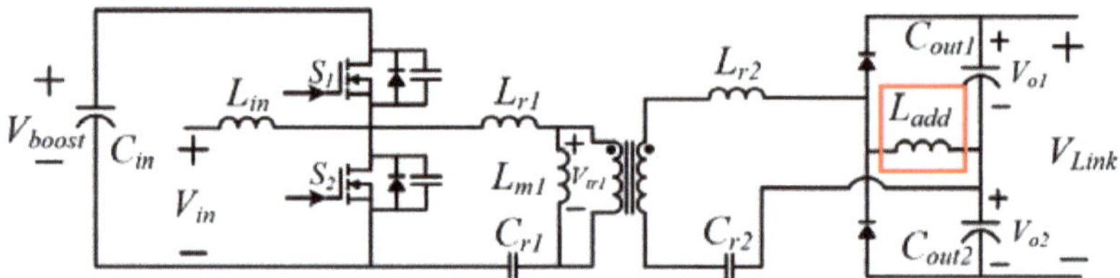

Figure 1. Schematic of the proposed CLLC resonant converter employing additional inductance.

Figure 2 shows the operational waveform of the conventional and the proposed current-fed CLLC resonant converter where S_1 and S_2 are the primary switches, $V_{ds,S1}$ and $V_{ds,S2}$ is the drain-source voltage of S_1 and S_2, respectively, I_{Lin} is the current passing through the input inductor, $I_{r,p}$ is the resonant current in the primary side, I_{Lm} is the magnetizing current, I_{S1} and I_{S2} are the currents passing through S_1 and S_2, respectively, and I_{Ladd} is the current in the additional inductance. The current-fed structure operates as the conventional boost converter by the switching operation of S_1 and S_2. The CLLC resonant tank employing the APWM has zero offset current on the magnetizing inductance which reduces the core and conduction losses of the transformer.

The current-fed structure is proper to compensate the wide input voltage variation. However, it limits the ZVS range according to the increase of the output load. The switch current of conventional voltage-fed CLLC resonant converter for the ZVS capability can be derived as follows:

$$ZVS_{S1} = I_r\left(t_{S2,off}\right) \tag{1}$$

$$ZVS_{S2} = -I_r\left(t_{S1,off}\right) \tag{2}$$

where $I_r(t_{s2,off})$ and $I_r(t_{s1,off})$ are the resonant current at the turn off state of power switches, respectively. The negative current of each switch is required to obtain the ZVS condition. In the voltage-fed CLLC resonant converter, the resonant current onlydetermines the ZVS condition of each switch. In the case of the current-fed CLLC resonant converter, the input inductor and resonant currents determine the ZVS current.The switch current of the current-fed CLLC resonant converter for the ZVS capability can be derived as follows:

$$ZVS_{S1,c} = I_r\left(t_{S2,off}\right) - I_{Lin}\left(t_{s2,off}\right) \tag{3}$$

$$ZVS_{S2,c} = -I_r\left(t_{S1,off}\right) + I_{Lin}\left(t_{s1,off}\right) \tag{4}$$

where $I_{Lin}(t_{s2,off})$ and $I_{Lin}(t_{s2,off})$ are the input inductor current at the turn off state of power switches, respectively. The high side switch (S_1) has a larger ZVS condition compared with Equation (1). However, the low side switch (S_2) has a poor ZVS condition compared with Equation (2). Figure 2a shows the theoretical waveforms of the conventional current-fed CLLC resonant converter, whichhas a hard switching operation on the S_2. The ZVS capability of the proposed CLLC resonant converter can be derived as follows:

$$ZVS_{S1,p} = I_r\left(t_{S2,off}\right) - I_{Lin}\left(t_{s2,off}\right) - nI_{Ladd}(t_{s2,off}) \tag{5}$$

$$ZVS_{S2,p} = -I_r\left(t_{S1,off}\right) + I_{Lin}\left(t_{s1,off}\right) - nI_{Ladd}\left(t_{s1,off}\right) \tag{6}$$

where $I_{Ladd}(t_{s2,off})$ and $I_{Ladd}(t_{s2,off})$ are the additional inductor currents at the turn off state of power switches, respectively, and n is the transformer turn ratio. Figure 2b shows the theoretical waveforms of the proposed CLLC resonant converter. The additional inductance extends the ZVS range compared with the conventional current-fed CLLC resonant converter.

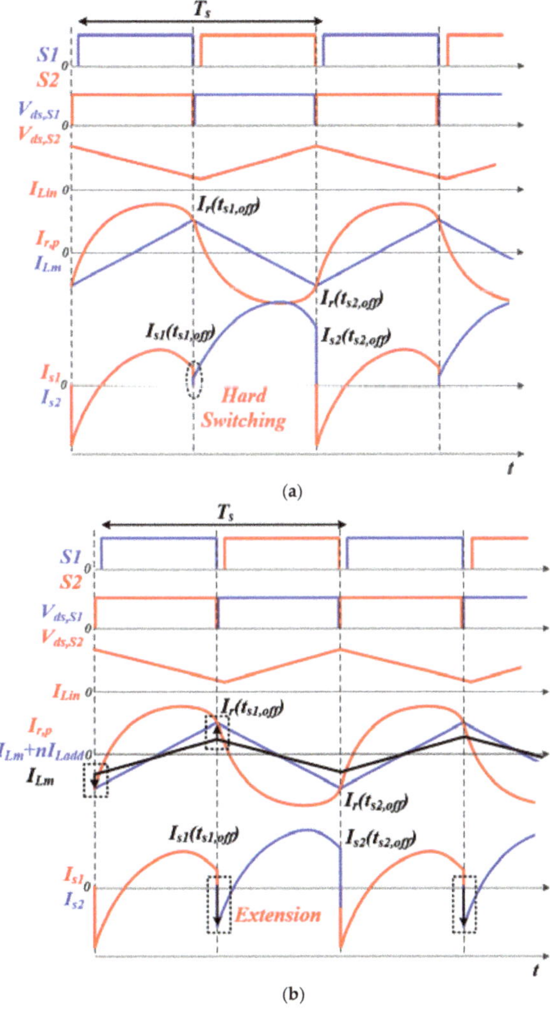

Figure 2. Theoretical operating waveforms: (**a**) Conventional current-fed CLLC resonant converter, (**b**) Proposed current-fed CLLC resonant converter employing additional inductance.

The turnon current for the ZVS condition of the current-fed CLLC resonant converter can be calculated with the input inductor current and resonant current. The ZVS condition on the low side switch can be derived as follows:

$$I_{zvs,s2} = \frac{1}{1 - D_{S2}} \frac{V_{boost}}{R} + \frac{(V_{in} - V_{boost})}{2L}(1 - D_{s2})T_s - \frac{V_{tr1}}{L_m} \frac{1 - D_{s2}}{2} T_s \tag{7}$$

where D_{s2} is the duty ratio of S_2, V_{boost} is the voltage of the current-fed structure, V_{in} is the input voltage of the battery, R is the load resistance, L is the input inductance, T_s is the switching time, L_m is the magnetizing inductance, and V_{tr1} is the transformer voltage. The negative value of $I_{zvs,S2}$ guarantees the ZVS capability of S_2. The decrease of the load resistance makes no ZVS condition of S_2. In addition, the large duty ratio of S_2 makes the worst ZVS condition, which means that the low input voltage condition is the worst ZVS condition.

The turn on current for ZVS condition of the low side switch with the proposed converter can be derived as follows:

$$I_{zvs,s2,p} = \frac{1}{1 - D_{s2}} \frac{V_{boost}}{R} + \frac{(V_{in} - V_{boost})}{2L}(1 - D_{s2})T_s - \frac{V_{tr1}}{L_m} \frac{1 - D_{s2}}{2} T_s - \frac{V_{o1}}{nL_{add}} \frac{1 - D_{s2}}{2} T_s \tag{8}$$

where V_{o1} is the voltage on the output capacitor as shown in Figure 1. The proposed converter extends the ZVS condition with the additional inductor on the secondary side as shown in Figure 2b. Figure 3 shows the ZVS capability according to the additional inductance. The ZVS current is the turn on current of S_2, which is required to have a negative value in order to obtain the ZVS capability. The small additional inductance is proper to extend the ZVS range. However, the small additional inductance increases the conduction loss on the primary side. Therefore, the design methodology of the additional inductance is required to obtain the ZVS capability for the entire load range.

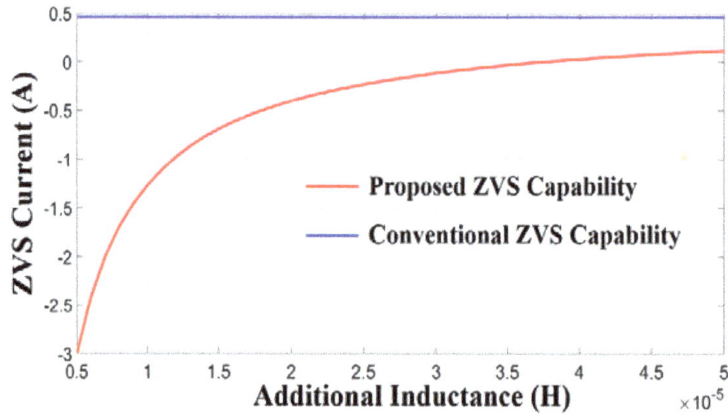

Figure 3. ZVS capability comparison between the conventional current-fed CLLC resonant converter and the proposed current-fed CLLC resonant converter.

The voltage gain according to the duty ratio can be described in Figure 4. The proposed converter has similar voltage gain to that of the conventional boost converter, which shows wide duty ratio variations to regulate the output voltage. The additional inductance can be designed at the worst duty ratio case, which is the maximum duty ratio of S_2.

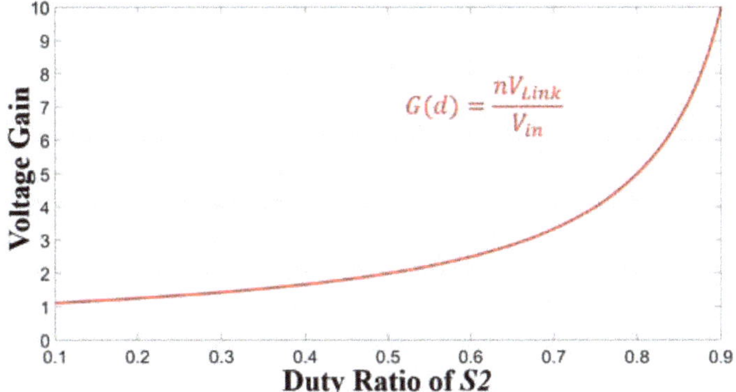

Figure 4. Gain according to duty ratio.

3. Design Methodology of Additional Inductance

The small additional inductance induces the conduction loss with large circulating current. The large additional inductance cannot obtain the ZVS capability of primary switches. Therefore, the maximum additional inductance is required to obtain the ZVS capability and the minimum conduction loss on the additional inductance, which can be derived with Equation (8) as follows:

$$L_{add} = \frac{V_{o1}}{nA}\frac{1}{2}\frac{D_{s2}}{2}T_s$$
$$A = \frac{1}{1-D_{s2}}\frac{V_{boost}}{R} + \frac{(V_{in}-V_{boost})}{2L}(1-D_{s2})T_s - \frac{V_{tr1}}{L_m}\frac{1-D_{s2}}{2}T_s \tag{9}$$

The proper additional inductance can be designed from Equation (9). Figure 5 shows the additional inductance according to the output load condition. The increment of output load requires smaller additional inductance. The design specification is shown in Table 1. The input voltage has a large variation by the state of charge (SOC) of the battery. The load voltage is fixed, and rated load is 20 W. The resonant frequency(f_r) is 200 kHz. The APWM requires the small resonant inductance to obtain the linear voltage gain according to the duty variation. The turn ratio is determined with the input voltage and output voltage ratio. The additional inductance is determined by (9).

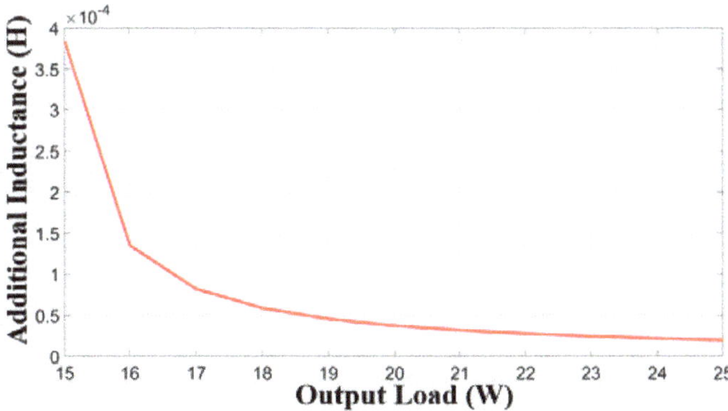

Figure 5. Desired additional inductance to obtain ZVS capability and minimum conduction loss according to output load condition.

Table 1. Design specification.

Parameter	Value
V_{in}	12 V–17 V
Load	32 V, 20 W
f_r	200 kHz
L_r	1 µH
L_m	30 µH
C_r	633 nF
Turn ratio	1:1
L_{add}	25 µH

4. Simulation and Experimental Results

The simulation results show the ZVS capability according to the additional inductance, as shown in Figure 6. The conventional current-fed CLLC resonant converter has no ZVS capability on the bottom side switch. However, the proposed CLLC resonant converter employing the additional inductance can achieve the ZVS condition for both the power switches.

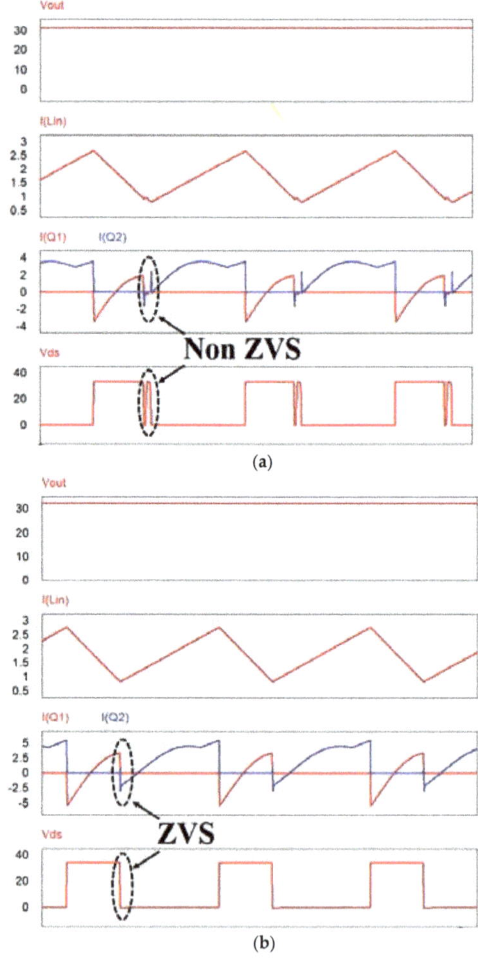

Figure 6. Simulation waveforms to verify ZVS capability: (**a**) Conventional current-fed CLLC resonant converter; (**b**) proposed current-fed CLLC resonant converter.

Figures 7–9 show the steady state waveforms of the conventional CLLC resonant converter according to the input voltage variation and load condition. The duty ratio regulates the output voltage according to the load conditions and input voltages.At the light load condition, it shows the ZVS operation. However, the CLLC resonant converter has a partial and no ZVS operation for the middle load and full load conditions, respectively. For allthe input voltage range, the ZVS can be achieved at only the light load condition.

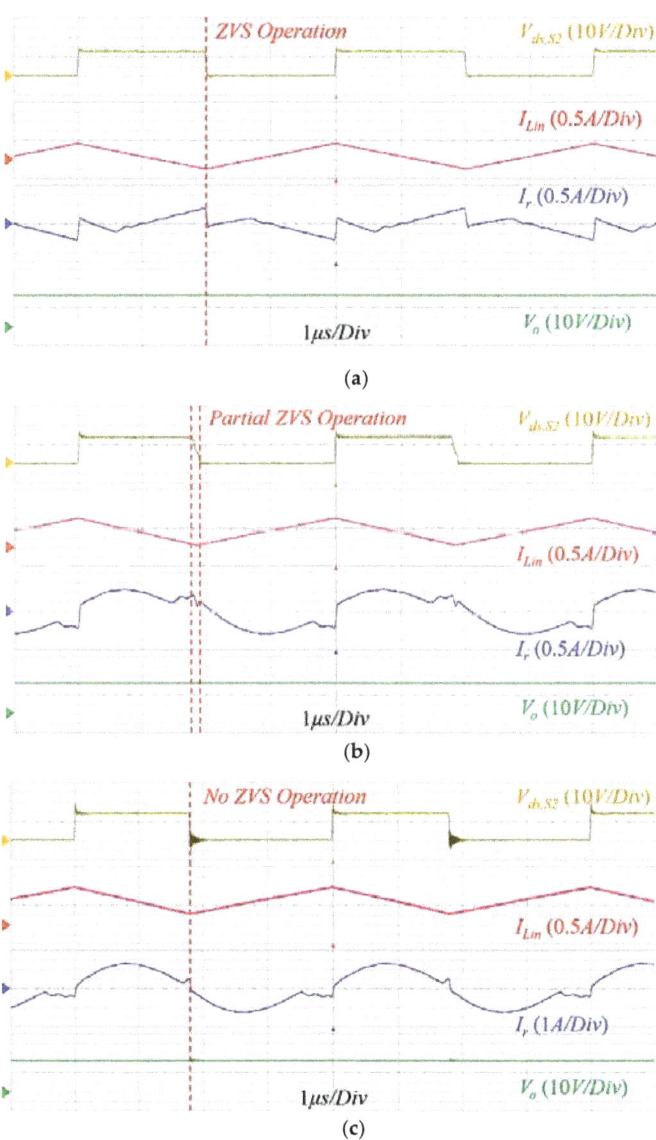

Figure 7. Experimental waveforms of the conventional CLLC resonant converter at 12 V condition: (**a**) 2 W light load condition; (**b**) 10 W middle load condition; (**c**) 20 W full load condition.

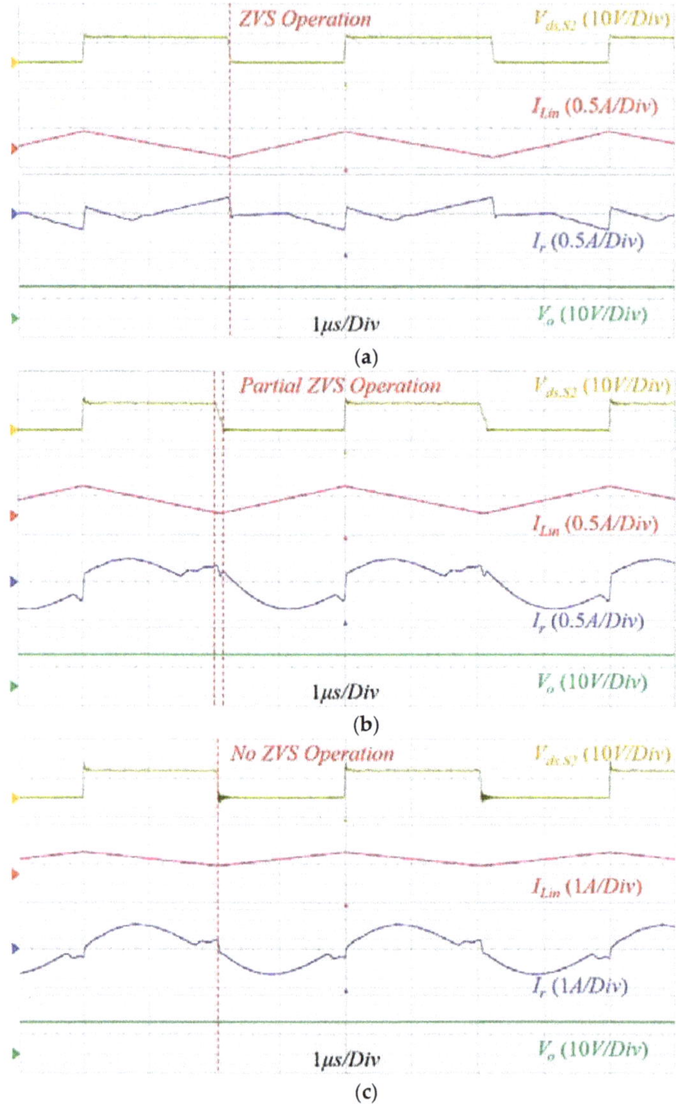

(a)

(b)

(c)

Figure 8. Experimental waveforms of the conventional CLLC resonant converter at 14 V condition: (a) 2 W light load condition; (b) 10 W middle load condition; (c) 20 W full load condition.

Figures 10–12 show the steady state waveforms of the proposed CLLC resonant converter according to the input voltages and load conditions.The proposed converter shows the ZVS operation for the entire load and input voltage conditions.The extended soft switching capability improves the power conversion efficiency compared with the partial or no ZVS cases. The power loss of the partial and no ZVS cases can be calculated as follows:

$$P_{noZVS} \cong \frac{1}{2}V_{ds,c}I_{on}t_{on}f_{sw} \tag{10}$$

where P_{noZVS} is the power loss according to the partial or no ZVS conditions, $V_{ds,c}$ is the drain-source voltage at the turn on state, I_{on} is the switch current at the turn on state, t_{on} is the turn on time duration, and f_{sw} is the switching frequency. Figure 13 shows the comparison of the power conversion efficiency between the conventional current-fed CLLC resonant converter and the proposed converter. For the light load condition, the proposed and conventional methods can obtain ZVS capability at the light load condition, which makes no difference in terms of the efficiency. However, the proposed converter has a higher power conversion efficiency for the middle to full load conditions. The maximum improvement of power conversion efficiency is 1% at the middle load condition.

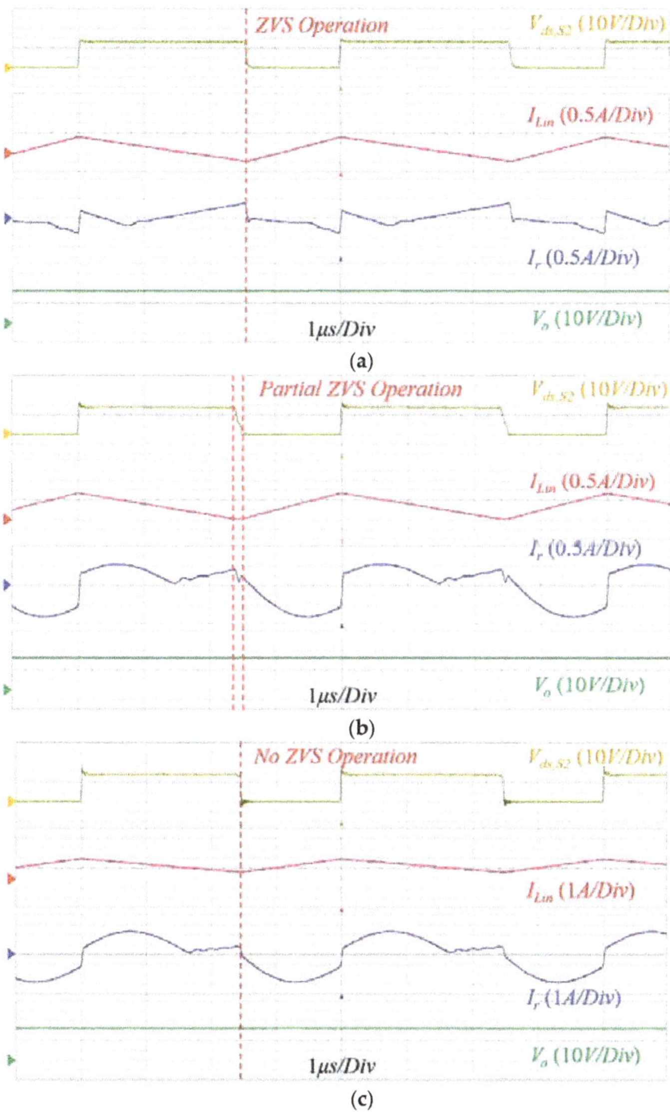

Figure 9. Experimental waveforms of the conventional CLLC resonant converter at 17 V condition: (**a**) 2 W light load condition; (**b**) 10 W middle load condition; (**c**) 20 W full load condition.

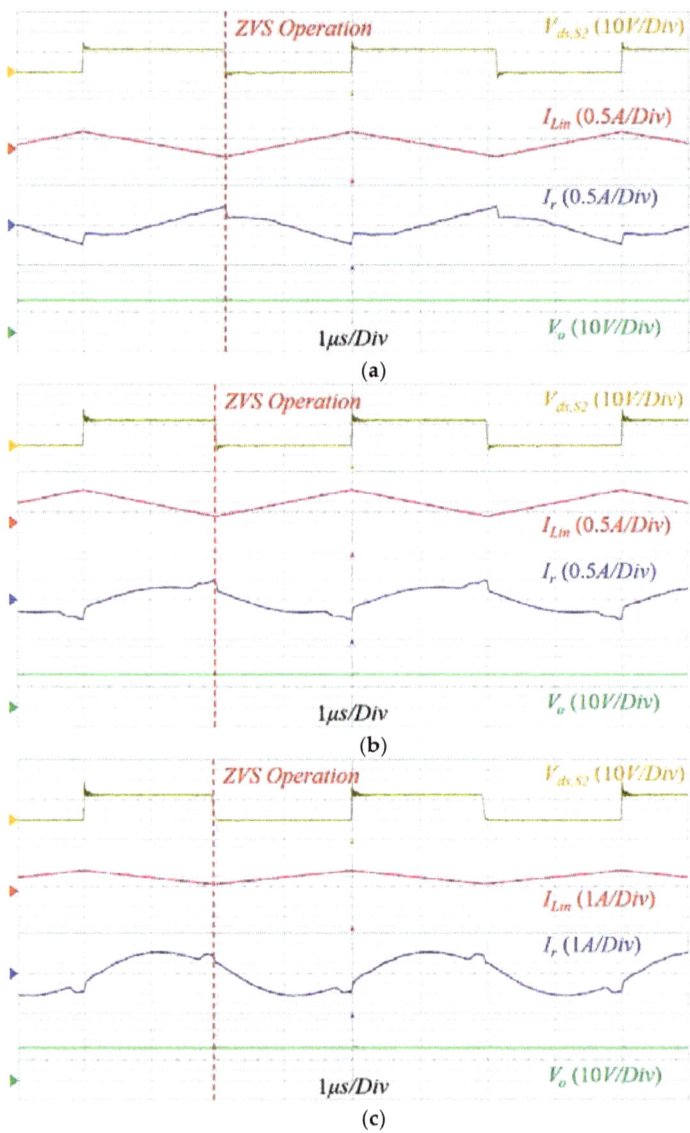

Figure 10. Experimental waveforms of the proposed CLLC resonant converter at 12 V condition: (a) 2 W light load condition; (b) 10 W middle load condition; (c) 20 W full load condition.

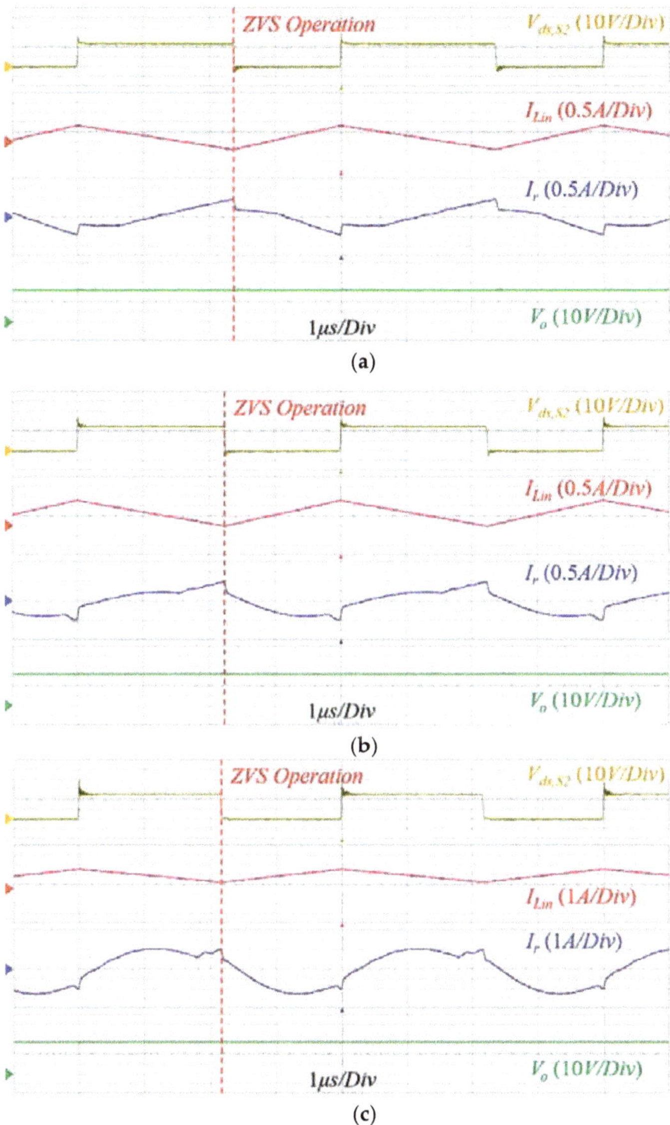

Figure 11. Experimental waveforms of the proposed CLLC resonant converter at 14 V condition: (**a**) 2 W light load condition; (**b**) 10 W middle load condition; (**c**) 20 W full load condition.

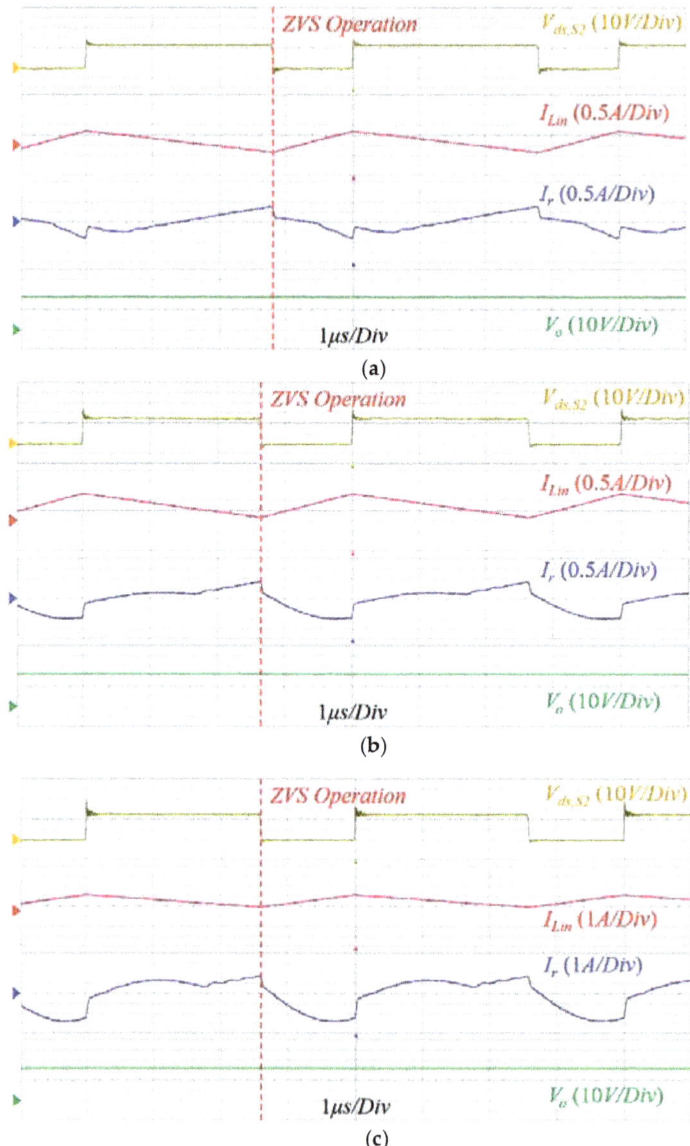

Figure 12. Experimental waveforms of the proposed CLLC resonant converter at 17 V condition: (**a**) 2 W light load condition; (**b**) 10 W middle load condition; (**c**) 20 W full load condition.

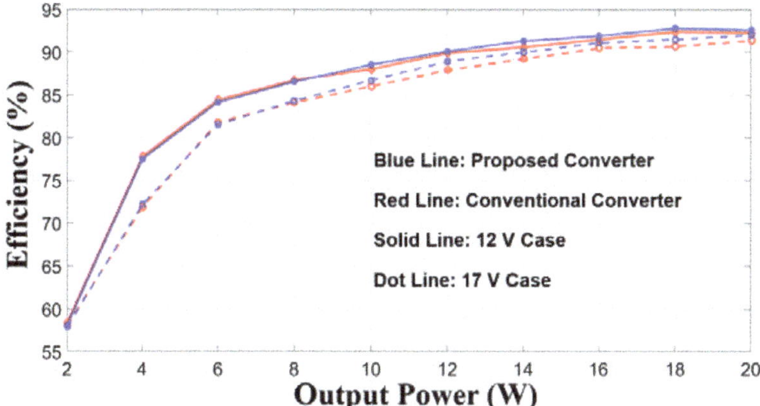

Figure 13. Power conversion efficiency curves.

5. Conclusions

A soft switching capability can improve the power conversion efficiency of the power converter. In this paper, the current-fed CLLC resonant converter employing the additional inductance is proposed to extend the ZVS capability. The operational principle and the design methodology of the additional inductance are analyzed to obtain the soft switching capability for the entire load and input voltage conditions. The simulation and experimental results verify the soft switching performance of the proposed CLLC resonant converter compared with the conventional converter. The maximum efficiency improvement using the proposed converter is 1% at the middle load condition.

Author Contributions: Conceptualization, H.P., J.J.; Methodology, H.P.; Software, H.P.; Validation, H.P.; Resources, D.K., S.B., J.J.; Writing-Original Draft Preparation, H.P.; Writing-Review & Editing, J.J.; Supervision, D.K, J.J.; Project Administration, D.K., S.B.; Funding Acquisition, J.J.

Funding: This research is performed based on the cooperation with UNIST-LIG Nex1 Cooperation.

Conflicts of Interest: The authors declare no conflict of interest.

References

1. De Doncker, R.W.; Divan, D.M.; Kheraluwala, M.H. A three-phase soft-switched high-power-density DC/DC converter for high-power applications. *IEEE Trans. Ind. Appl.* **1991**, *27*, 63–73. [CrossRef]
2. Liu, Y.-C.; Chen, M.-C.; Yang, C.-Y.; Kim, K.A.; Chiu, H.-J. High-Efficiency Isolated Photovoltaic Microinverter Using Wide-Band Gap Switches for Standalone and Grid-Tied Applications. *Energies* **2018**, *11*, 569. [CrossRef]
3. Zaman, H.; Wu, X.; Zheng, X.; Khan, S.; Ali, H. Suppression of Switching Crosstalk and Voltage Oscillations in a SiC MOSFET Based Half-Bridge Converter. *Energies* **2018**, *11*, 3111. [CrossRef]
4. Pilawa-Podgurski, R.C.N.; Sagneri, A.D.; Rivas, J.M.; Anderson, D.I.; Perreault, D.J. Very-High-Frequency Resonant Boost Converters. *IEEE Trans. Power Electron.* **2009**, *24*, 1654–1665. [CrossRef]
5. Reusch, D.; Strydom, J. Understanding the Effect of PCB Layout on Circuit Performance in a High-Frequency Gallium-Nitride-Based Point of Load Converter. *IEEE Trans. Power Electron.* **2014**, *29*, 2008–2015. [CrossRef]
6. Kaiwei, Y.; Yang, Q.; Ming, X.; Lee, F.C. A novel winding-coupled buck converter for high-frequency, high-step-down DC-DC conversion. *IEEE Trans. Power Electron.* **2005**, *20*, 1017–1024.
7. Meng, Z.; Wang, Y.-F.; Yang, L.; Li, W. High Frequency Dual-Buck Full-Bridge Inverter Utilizing a Dual-Core MCU and Parallel Algorithm for Renewable Energy Applications. *Energies* **2017**, *10*, 402. [CrossRef]
8. Huang, R.; Mazumder, S.K. A Soft-Switching Scheme for an Isolated DC/DC Converter With Pulsating DC Output for a Three-Phase High-Frequency-Link PWM Converter. *IEEE Trans. Power Electron.* **2009**, *24*, 2276–2288. [CrossRef]

9. Liu, X.; Liu, J.; Wang, J.; Wang, C.; Yuan, X. Design Method for the Coil-System and the Soft Switching Technology for High-Frequency and High-Efficiency Wireless Power Transfer Systems. *Energies* **2018**, *11*, 7. [CrossRef]

10. Liu, K.H.; Lee, F.C. Zero-voltage switching technique in DC/DC converters. *IEEE Trans. Power Electron.* **1990**, *5*, 293–304. [CrossRef]

11. Lin, B.-R.; Wu, G.-Y. Bidirectional Resonant Converter with Half-Bridge Circuits: Analysis, Design, and Implementation. *Energies* **2018**, *11*, 1259. [CrossRef]

12. Lin, B.-R. Investigation of a Resonant dc–dc Converter for Light Rail Transportation Applications. *Energies* **2018**, *11*, 1078. [CrossRef]

13. Zhang, S.-H.; Luo, F.-Z.; Wang, Y.-F.; Liu, J.-H.; He, Y.-P.; Dong, Y. Control Method Based on Demand Response Needs of Isolated Bus Regulation with Series-Resonant Converters for Residential Photovoltaic Systems. *Energies* **2017**, *10*, 752. [CrossRef]

14. Lin, R.; Huang, L. Efficiency Improvement on LLC Resonant Converter Using Integrated LCLC Resonant Transformer. *IEEE Trans. Ind. Appl.* **2018**, *54*, 1756–1764. [CrossRef]

15. Tan, X.; Ruan, X. Equivalence Relations of Resonant Tanks: A New Perspective for Selection and Design of Resonant Converters. *IEEE Trans. Ind. Electron.* **2016**, *63*, 2111–2123. [CrossRef]

16. Fayyaz, A.; Romano, G.; Urresti, J.; Riccio, M.; Castellazzi, A.; Irace, A.; Wright, N. A Comprehensive Study on the Avalanche Breakdown Robustness of Silicon Carbide Power MOSFETs. *Energies* **2017**, *10*, 452. [CrossRef]

17. Efthymiou, L.; Camuso, G.; Longobardi, G.; Chien, T.; Chen, M.; Udrea, F. On the Source of Oscillatory Behaviour during Switching of Power Enhancement Mode GaN HEMTs. *Energies* **2017**, *10*, 407. [CrossRef]

18. Seeman, M.D. GaN Devices in Resonant LLC Converters: System-level considerations. *IEEE Power Electron. Mag.* **2015**, *2*, 36–41. [CrossRef]

19. Zhang, W.; Wang, F.; Costinett, D.J.; Tolbert, L.M.; Blalock, B.J. Investigation of Gallium Nitride Devices in High-Frequency LLC Resonant Converters. *IEEE Trans. Power Electron.* **2017**, *32*, 571–583. [CrossRef]

20. Chen, W.; Rong, P.; Lu, Z. Snubberless Bidirectional DC–DC Converter with New CLLC Resonant Tank Featuring Minimized Switching Loss. *IEEE Trans. Ind. Electron.* **2010**, *57*, 3075–3086. [CrossRef]

21. Ryu, M.; Kim, H.; Baek, J.; Kim, H.; Jung, J. Effective Test Bed of 380-V DC Distribution System Using Isolated Power Converters. *IEEE Trans. Ind. Electron.* **2015**, *62*, 4525–4536. [CrossRef]

22. He, P.; Khaligh, A. Comprehensive Analyses and Comparison of 1 kW Isolated DC–DC Converters for Bidirectional EV Charging Systems. *IEEE Trans. Transp. Electrif.* **2017**, *3*, 147–156. [CrossRef]

23. Zou, S.; Lu, J.; Mallik, A.; Khaligh, A. Bi-Directional CLLC Converter With Synchronous Rectification for Plug-In Electric Vehicles. *IEEE Trans. Ind. Appl.* **2018**, *54*, 998–1005. [CrossRef]

24. Jung, J.; Kim, H.; Ryu, M.; Baek, J. Design Methodology of Bidirectional CLLC Resonant Converter for High-Frequency Isolation of DC Distribution Systems. *IEEE Trans. Power Electron.* **2013**, *28*, 1741–1755. [CrossRef]

25. Zhang, C.; Li, P.; Kan, Z.; Chai, X.; Guo, X. Integrated Half-Bridge CLLC Bidirectional Converter for Energy Storage Systems. *IEEE Trans. Ind. Electron.* **2018**, *65*, 3879–3889. [CrossRef]

26. Sun, X.; Shen, Y.; Zhu, Y.; Guo, X. Interleaved Boost-Integrated LLC Resonant Converter with Fixed-Frequency PWM Control for Renewable Energy Generation Applications. *IEEE Trans. Power Electron.* **2015**, *30*, 4312–4326. [CrossRef]

Article

High-Frequency LLC Resonant Converter with GaN Devices and Integrated Magnetics

Yu-Chen Liu [1,*], Chen Chen [2], Kai-De Chen [2], Yong-Long Syu [2] and Meng-Chi Tsai [1]

[1] Department of Electrical Engineering, National Ilan University, No. 1 Section 1 Shennong Road, Yilan 260, Taiwan; kid3181910@gmail.com

[2] Department of Electronic Engineering, National Taiwan University of Science, and Technology, No. 43 Section 4 Keelung Rd Da'an District, Taipei 10607, Taiwan; D10502204@mail.ntust.edu.tw (C.C.); d10502203@mail.ntust.edu.tw (K.-D.C.); x452800@gmail.com (Y.-L.S.)

* Correspondence: ycliu@niu.edu.tw

Received: 15 April 2019; Accepted: 5 May 2019; Published: 10 May 2019

Abstract: In this study, a light emitting diode (LED) driver containing an integrated transformer with adjustable leakage inductance in a high-frequency isolated LLC resonant converter was proposed as an LED lighting power converter. The primary- and secondary-side topological structures were analyzed from the perspectives of component loss and component stress, and a full-bridge structure was selected for both the primary- and secondary-side circuit architecture of the LLC resonant converter. Additionally, to achieve high power density and high efficiency, adjustable leakage inductance was achieved through an additional reluctance length, and the added resonant inductor was replaced with the transformer leakage inductance without increasing the amount of loss caused by the proximity effect. To optimize the transformer, the number of primary- and secondary-side windings that resulted in the lowest core loss and copper loss was selected, and the feasibility of the new core design was verified using ANSYS Maxwell software. Finally, this paper proposes an integrated transformer without any additional resonant inductor in the LLC resonant converter. Transformer loss is optimized by adjusting parameters of the core structure and the winding arrangement. An LLC resonant converter with a 400 V input voltage, 300 V output voltage, 1 kW output power, and 500 kHz switching frequency was created, and a maximum efficiency of 97.03% was achieved. The component with the highest temperature was the transformer winding, which reached 78.6 °C at full load.

Keywords: LLC resonant converter; integrated transformer; adjustable leakage inductance; LED driver

---

1. Introduction

Light emitting diode (LED) lighting has become more popular in recent years, particularly as product energy efficiency has become more important to consumers [1–3]. Compared with conventional lighting equipment, LEDs are brighter and have a longer service life. LEDs have thus been used in various lighting scenarios, such as street lighting, indoor lighting, and backlighting. They have also been used for lighting equipment with high power requirements, like baseball field or basketball court lights. Most of these LED driver circuits have a two-stage architecture consisting of a power factor correction converter in the first stage and an isolated buck converter in the second stage, with an LLC resonant converter usually employed as the second-stage circuit [4,5].

Unlike other common isolated buck converters, an LLC resonant converter is characterized by power components with zero voltage switching (ZVS) and zero current switching at the high—and low—voltage ends, respectively, in a full load range. Additionally, unlike phase-shifted full-bridge converters, an LLC converter has no output inductance at the low-voltage end and thus has higher conversion efficiency [6]. Moreover, to achieve high power density, increasing the switching frequency of the power switch reduces the volume of the magnetic component; however, this is accompanied by a considerable increase in switching loss. Therefore, in addition to the ZVS feature of the LLC resonant converter, wide-band-gap switch components should also be employed to reduce the switching loss under high-frequency operation [7,8]. Additionally, selecting an appropriate LLC resonant converter architecture can further optimize the efficiency and reduce component stress [9–11]. The topologies commonly used in the high-voltage end of the LLC resonant converter are the half bridge and full bridge, whereas those used in the low-voltage end are the full bridge or center tap.

To achieve high power density, the leakage inductance of the transformer in this study was used as the resonant inductor of the circuit [12–14], which reduces the number of magnetic components required. Previous research replaced the resonant inductor with the leakage inductance of the transformers, mostly through winding the transformer using litz wire and setting the operating frequency between 50 and 200 kHz. However, when the operating frequency is between 500 kHz and 1 MHz, the litz wire is usually replaced by printed circuit board (PCB) windings to reduce the alternating-current (AC) resistance of the copper traces under high-frequency operation. Additionally, the primary- and secondary-side windings are interleaved to weaken the proximity effect. This approach reduces proximity-effect-induced loss and minimizes the leakage inductance of the transformer [15–17], which is a disadvantage in LLC resonant converters that require leakage inductance to design resonant tanks. A conventional method for increasing the leakage inductance is to increase the spacing between the primary and secondary sides [18–20], which is mostly achieved using a non-interleaved winding that leads to a sharp rise in the proximity-effect-induced AC loss. Therefore, the concept of adjustable leakage inductance [21,22] was adopted in this study, with an additional reluctance length added between the primary-side and secondary-side windings of the transformer. This enabled the magnetic flux generated by the primary-side winding to flow into the added reluctance length, thereby increasing the primary-side leakage inductance. The proposed approach weakens the proximity effect in the interleaved winding and attains the objective of increasing the leakage inductance. Compared to existing LED power converter designs [23–25], this work reduces the switching loss via wide bandgap devices and soft switching techniques under 500 kHz switching frequency conditions and utilizes an integrated transformer without an additional resonant inductor to achieve high power density. Ultimately, an LLC resonant converter with an output power of 1 kW, switching frequency of 500 kHz, input voltage of 400 V, and output voltage of 300 V was achieved in this study.

In this study, a second-stage LLC resonant converter was proposed for the LED driver circuit, and the LLC resonant circuit was used as a direct-current transformer (DCX) to operate the driver circuit at high efficiency. In addition, this study also designed and optimized the primary- and secondary-side topologies and magnetic components in the LLC resonant converter. In Section 2, under the conditions of 1 kW output power, 400 V input voltage, and 300 V output voltage, the loss and component stress at the primary—and secondary—side structures of the converter are analyzed and a topology suitable for the specification in this study is selected. Section 3 details the mathematical induction of adjustable leakage inductance and the appropriate number of primary- and secondary-side windings for the new core based on the optimal point of core loss and copper loss. The feasibility of the designed core structure was verified using ANSYS Maxwell software (ANSYS, Canonsburg, PA, USA). Section 4 presents the experimental results, including temperature distribution maps.

2. LLC Topology Comparison

The primary and secondary sides of different architectures are discussed in this section. Because different circuit specifications have their appropriate architecture, the architecture is chosen based on power level and component stress, as shown in Figure 1. The primary side may consist of a half bridge or full bridge, whereas the secondary side comprises a center-tap and full-bridge rectification. The choice of different architectures for the primary side results in different transformer designs; hence, the primary side is mainly selected based on the corresponding architecture for the appropriate power level. As for the different architectures on the secondary side, there are different component stresses and numbers of switches. Therefore, the appropriate architecture for the secondary side is mainly selected according to the level of the output voltage and current.

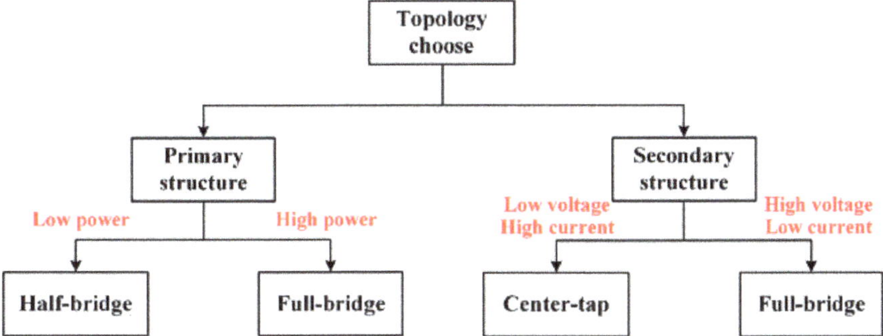

Figure 1. Architecture selection diagram.

The common structures for the primary side are the half bridge and full bridge (Figure 2). For the half bridge, the voltage across the resonant tank is $+V_{in}$ to 0 V; however, the actual voltage across the transformer is $\pm V_{in}/2$. Therefore, the turns ratio between the primary and secondary sides must be accounted for when designing the transformer. The energy transfer formula in Equation (1) can be rewritten as a general formula that establishes ZVS under a half-bridge structure, as shown in Equation (2).

$$\frac{1}{2}LI^2 \geq \frac{1}{2}CV^2 \tag{1}$$

in which the unit of L is Henry (H), I is Ampere (A), C is Farad (F), and V is Volt (V).

$$L_m \leq \frac{t_{dead}}{16 C_{oss} f_{sw}} \tag{2}$$

in which the unit of L_m is Henry (H), t_{dead} is seconds (s), C_{oss} is Farad (F), and f_{sw} is Hertz (Hz).

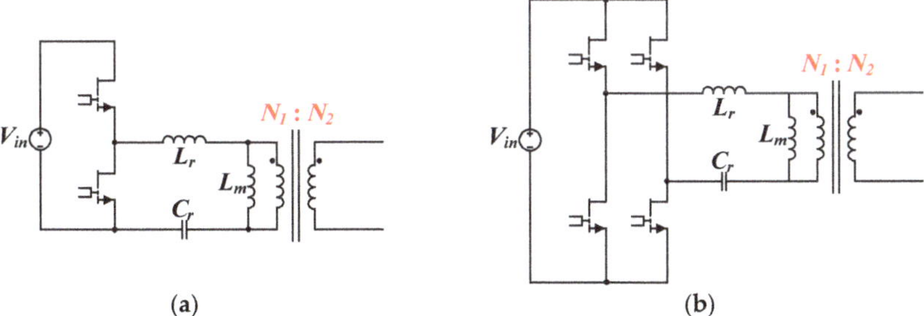

Figure 2. Primary structure (**a**) half bridge and (**b**) Full bridge topology.

For a full-bridge structure, the voltage across the transformer is approximately $\pm V_{in}$, and the input voltage can be directly divided by the output voltage to obtain the turns ratio when designing the transformer. For primary-side switching to achieve ZVS in a full-bridge structure, the energy transfer formula Equation (1) can be rewritten as the following general formula:

$$L_m \leq \frac{t_{dead}}{8 C_{oss} f_{sw}} \tag{3}$$

Under the conditions of 400 V input voltage, 300 V output voltage, and a half-bridge structure of the primary side, the equivalent turns ratio of the primary and secondary sides is indicated by Equation (4), whereas Equation (5) obtains the equivalent turns ratio of the two sides when a full-bridge structure is adopted on the primary side. In choosing the primary-side architecture, the conditions of primary-side ZVS are influenced by the stray capacitance [26]. In particular, the stray capacitance of the secondary side mapped back to the primary side is affected by the turns ratio. Thus, if a half-bridge structure was adopted under this specification, the stray capacitance mapped from the secondary to the primary side would be magnified. Therefore, a full-bridge structure was employed in this study to reduce the effect on ZVS condition. Table 1 [27] shows a comparison of the various ratings of the two primary-side architectures.

$$\frac{N_1}{N_2} = \frac{V_{in}/2}{V_o} = \frac{2}{3} \tag{4}$$

$$\frac{N_1}{N_2} = \frac{V_{in}}{V_o} = \frac{4}{3} \tag{5}$$

Table 1. Comparison of primary-side architectures.

Primary Structure	Half-Bridge	Full-Bridge
Primary device voltage stress	V_{in} (400 V)	V_{in} (400 V)
Primary device current stress	$\dfrac{V_o\sqrt{4\pi^2+\left(\frac{N_1}{N_2}\right)^4 R_L{}^2\left(\frac{T}{L_m}\right)^2}}{4\cdot\frac{N_1}{N_2}\cdot R_L}$ (8.25 A)	$\dfrac{V_o\sqrt{4\pi^2+\left(\frac{N_1}{N_2}\right)^4 R_L{}^2\left(\frac{T}{L_m}\right)^2}}{4\cdot\frac{N_1}{N_2}\cdot R_L}$ (4.81 A)
Primary device rms current	$\dfrac{V_o\sqrt{4\pi^2+\left(\frac{N_1}{N_2}\right)^4 R_L{}^2\left(\frac{T}{L_m}\right)^2}}{8\cdot\frac{N_1}{N_2}\cdot R_L}$ (4.08 A)	$\dfrac{V_o\sqrt{4\pi^2+\left(\frac{N_1}{N_2}\right)^4 R_L{}^2\left(\frac{T}{L_m}\right)^2}}{8\cdot\frac{N_1}{N_2}\cdot R_L}$ (2.36 A)
Resonant inductor rms current	$\dfrac{V_o\sqrt{4\pi^2+\left(\frac{N_1}{N_2}\right)^4 R_L{}^2\left(\frac{T}{L_m}\right)^2}}{4\sqrt{2}\cdot\frac{N_1}{N_2}\cdot R_L}$ (5.8 A)	$\dfrac{V_o\sqrt{4\pi^2+\left(\frac{N_1}{N_2}\right)^4 R_L{}^2\left(\frac{T}{L_m}\right)^2}}{4\sqrt{2}\cdot\frac{N_1}{N_2}\cdot R_L}$ (3.38 A)
ZVS condition	$L_m \leq \dfrac{t_{dead}}{16 C_{oss} f_{sw}}$	$L_m \leq \dfrac{t_{dead}}{8 C_{oss} f_{sw}}$
Turns ratio	$N_1{:}N_2$ (12:18)	$N_1{:}N_2$ (24:18)
Primary device	GS66508B, 650 V	GS66508B, 650 V
Number of primary device	2	4
Total SR conduction loss	$I_{rms}{}^2\cdot R_{ds(on)}\cdot 2$ (1.665 W)	$I_{rms}{}^2\cdot R_{ds(on)}\cdot 4$ (1.114 W)
Winding loss of primary side	7.24 W	5.51 W
Proper specification	Low power	High power

In Table 1, T is the switching period, and its unit is seconds (s), and RL is the output load. Its unit is Ohm (Ω). ZVS, zero voltage switching. SR, synchronous rectifier.

Common structures for the secondary side are center-tap and full-bridge rectification (Figure 3). For center-tap rectification, the voltage over the secondary side of the transformer was $1 \times V_o$; however, the voltage over the synchronous rectifiers (SRs) was $2 \times V_o$, and overall, the conduction loss amounted to that of two SRs. Based on this analysis, the center-tap structure was more suitable for low-output-voltage and high-current specifications.

Compared with those in the center-tap structure, the voltage over the secondary side and the SRs of the transformer in the full-bridge rectification structure were $1 \times V_o$; however, the conduction loss amounted to that of four SRs. Based on this analysis, full-bridge rectification was suitable for high-output-voltage and low-current specifications. For the high-output-voltage specification and considering the tradeoff between the SR voltage rating, SR conduction loss, and secondary-side copper loss, full wave rectification was selected for the secondary-side structure, and the switch employed was the GS66508B. Table 2 shows a comparison of the SR rated voltage, transformer turns ratio, and SR number of the two secondary-side architectures.

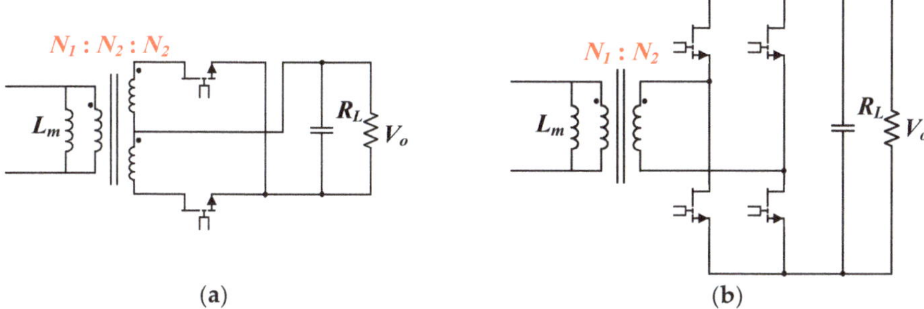

Figure 3. Secondary-side architecture: (**a**) center tap and (**b**) full bridge.

Table 2. Comparison of secondary-side architectures.

Secondary Structure	Center-Tap	Full-Bridge
SR voltage stress	$2\,V_0$ (600 V)	V_0 (300 V)
SR current stress	$\sqrt{12}\,\dfrac{V_o\sqrt{12\pi^4 + \frac{5\pi^2-48}{L_m^2}\left(\frac{N_1}{N_2}\right)^4 R_L{}^2 T^2}}{24\pi R_L}$ (5.5 A)	$\sqrt{12}\,\dfrac{V_o\sqrt{12\pi^4 + \frac{5\pi^2-48}{L_m^2}\left(\frac{N_1}{N_2}\right)^4 R_L{}^2 T^2}}{24\pi R_L}$ (5.5 A)
SR rms current	$\sqrt{3}\,\dfrac{V_o\sqrt{12\pi^4 + \frac{5\pi^2-48}{L_m^2}\left(\frac{N_1}{N_2}\right)^4 R_L{}^2 T^2}}{24\pi R_L}$ (2.67 A)	$\sqrt{3}\,\dfrac{V_o\sqrt{12\pi^4 + \frac{5\pi^2-48}{L_m^2}\left(\frac{N_1}{N_2}\right)^4 R_I{}^2 T^2}}{24\pi R_L}$ (2.67 A)
Turns ratio	N1:N2:N2 (24:18:18)	N1:N2 (24:18)
SR device	GS66508B, 650 V	GS66508B, 650 V
Number of SR device	2	4
Total SR conduction loss	$I_{rms}{}^2 \cdot R_{ds(on)} \cdot 2$ (0.713 W)	$I_{rms}{}^2 \cdot R_{ds(on)} \cdot 4$ (1.426 W)
Winding loss of secondary side	2.41 W	2.41 W
Proper specification	Low output voltage High output current	High output voltage Low output current

Next, the differences between the four architectures (primary and secondary side) are compared. In terms of the primary-side switch voltage stress, both the half bridge and full bridge result in a voltage stress of $1 \times V_{in}$, whereas the secondary-side switch voltage stress is $2\,V_0$ and V_0 for center-tap and full-bridge rectification, respectively. The voltage over the transformer was $\pm 0.5\,V_{in}$ when a half-bridge structure was employed for the primary side, whereas using a full-bridge structure yielded a voltage of $\pm V_{in}$. Therefore, when transmitting the same power, the transformer current was I_r when a full-bridge structure was used on the primary side but had to be $2\,I_r$ if a half-bridge structure was used. Because the voltage over the transformer was half the input voltage, the current on the primary side was greater when a half-bridge structure was employed than when the primary side comprised a full-bridge structure. Therefore, the primary-side half-bridge architecture is suitable for circuits with low power applications to prevent excessive current flow through the transformer, which would result in saturation. Regarding the secondary side, the component voltage stress when the center-tap structure was used was twice that when full-bridge rectification was employed, whereas the number of components was half; the center-tap structure is thus appropriate for low-voltage-output and high-current applications. Table 3 shows a comparison of the characteristics for the four architectures.

Table 3. Comparison of the four architectures.

Component	Primary Side: Half-Bridge Secondary Side: Center-Tap	Primary Side: Half-Bridge Secondary Side: Full-Bridge	Primary Side: Full-bridge Secondary side: Center-Tap	Primary side: Full-Bridge Secondary side: Full-Bridge
Primary device voltage stress	V_{in}	V_{in}	V_{in}	V_{in}
SR voltage stress	$2\,V_0$	V_0	$2V_0$	V_0
Primary-side transformer voltage stress	$\pm 0.5\,V_{in}$	$\pm 0.5\,V_{in}$	$\pm V_{in}$	$\pm V_{in}$
Secondary-side transformer voltage stress	$\pm V_0$	$\pm V_0$	$\pm V_0$	$\pm V_0$
Primary-side transformer current	$2\,I_r$	$2\,I_r$	I_r	I_r
Proper specification	Low power Low output voltage High output current	Low power High output voltage Low output current	High power Low output voltage High output current	High power High output voltage Low output current

The loss distribution of the four architectures was calculated on the basis of the specifications in this study (Figure 4). According to the loss distribution illustrated in Figure 4, the loss of each architecture type was mostly copper loss of the transformers and other loss types, including the equivalent series resistance of the capacitor and via loss. The overall efficiency of the architecture comprising a primary-side full-bridge structure and secondary-side center-tap structure was the highest of all architectures. However, because the specification of this study was high output voltage, the SR voltage rating of the secondary-side center-tap structure had to be 2 V_o; additionally, the actual test circuit was not ideal. Therefore, the architecture with primary-side full-bridge and secondary-side full bridge structures, which had the second highest efficiency, was selected to prevent the SR from being damaged by the high voltage.

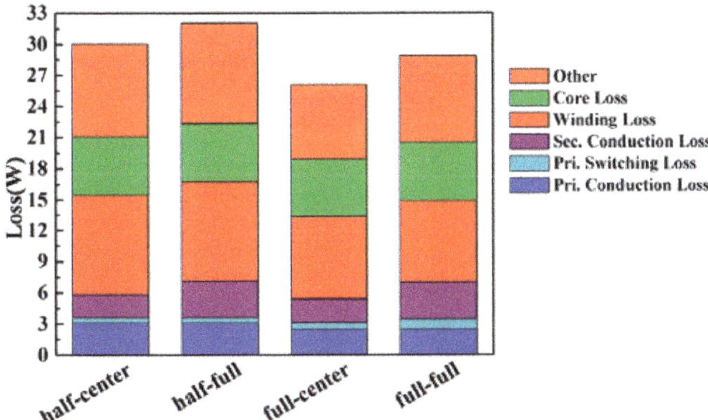

Figure 4. Comparison of loss types in different architectures.

3. Design Concept and Loss Optimization of Adjustable Leakage Inductance

The conventional integrated leakage inductance is formed by leakage flux that is lost to the air and does not pass through the secondary side. Accurate control of the amount of leakage inductance is practically difficult. The concept of adjustable leakage inductance that was adopted in this study was to add an additional reluctance length that prevents some flux generated on the primary side from flowing into the secondary-side winding, in turn generating leakage inductance. The leakage inductance used in this study was designed using the concept of split flow of the transformer reluctance.

Figure 5 presents the concept of adjustable leakage inductance, where the voltage of the primary side and secondary side are denoted by Vp and Vs, respectively. The winding in the primary side was divided into two parts, of which one part wound counterclockwise with 18 turns around the left outer leg, and the other wound clockwise with 6 turns around the right outer leg. Similarly, the winding on the secondary side was also divided into two parts, in which one part wound clockwise with 6 turns around the left outer leg, and the other wound counterclockwise with 12 turns around the right outer leg. Moreover, according to Equation (6), the magnetomotive force (MMF) of these windings could be represented as 18 I_p, 6 I_s, 6 I_p, and 12 I_s, and their voltage as V_{pw}, V_{px}, V_{sy}, and V_{sz}, respectively. R_o represents the reluctance of the outer leg, and R_c represents the reluctance of the center leg.

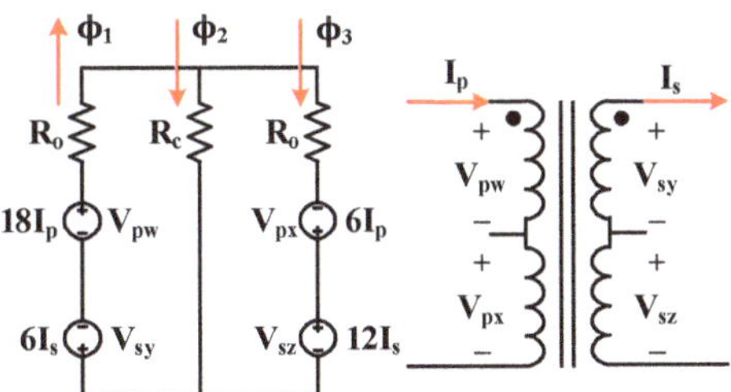

Figure 5. Reluctance model for adjustable leakage inductance.

$$\phi = \frac{mmf}{R} = NI \qquad (6)$$

Figure 6 displays the equivalence of the reluctance generated by the passing flux produced in the left leg that was achieved by using the superposition theorem. Regarding the core structure, because the reluctances of both outer legs are the same, the equivalent reluctance passing through ϕ_1 and ϕ_3 are the same, as indicated in Equation (7). After the reluctance passing through the flux has been obtained,

ϕ_1 could be presented as Equation (8); although the equivalent reluctance *of* ϕ_1 and ϕ_{3} are the same, ϕ_3 could be presented as in Equation (9) because the MMF is different for each leg. Because the center leg had no windings, only the values of flux on the left and right outer legs are listed.

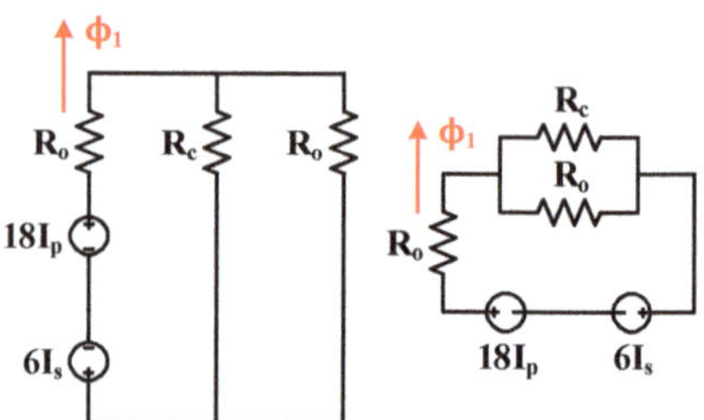

Figure 6. Equivalent reluctance model for flux split flow.

$$R_{\phi 1} = R_{\phi 3} = \frac{R_o(R_o + 2R_c)}{R_o + R_c} \tag{7}$$

$$\phi_1 = \left(18I_p - 6I_s\right) \cdot \frac{R_o + R_c}{R_o(R_o + 2R_c)} \tag{8}$$

$$\phi_3 = \left(6I_p - 12I_s\right) \cdot \frac{R_o + R_c}{R_o(R_o + 2R_c)} \tag{9}$$

As mentioned, the MMF generated from the windings on each leg was determined using the superposition theorem. Subsequently, Equations (10)–(12) represent the total flux of each leg in consideration of flux split flow.

$$\phi_{1\_total} = \frac{18R_o + 24R_c}{R_o(R_o + 2R_c)} \cdot I_p - \frac{6R_o + 18R_c}{R_o(R_o + 2R_c)} \cdot I_s \tag{10}$$

$$\phi_{2\_total} = \frac{12}{R_o(R_o + 2R_c)} \cdot I_p - \frac{6}{R_o(R_o + 2R_c)} \cdot I_s \tag{11}$$

$$\phi_{3\_total} = \frac{6R_o + 24R_c}{R_o(R_o + 2R_c)} \cdot I_p - \frac{12R_o + 18R_c}{R_o(R_o + 2R_c)} \cdot I_s \tag{12}$$

According to Figure 7, the primary-side voltage of the transformer V_p is the sum of V_{pw} and V_{px}; the secondary-side voltage of the transformer V_s is the sum of V_{sy} and V_{sz}. As demonstrated by Equation (13), Faraday's law of induction was employed to ascertain the relationship between the voltage, number of turns, and variation in flux over time. Moreover, the primary-side voltage V_p and secondary-side voltage V_s of the transformer are represented using Equations (14) and (15).

$$V = n\frac{d\phi}{dt} \tag{13}$$

$$V_p = \frac{360R_o + 576R_c}{R_o(R_o + 2R_c)} \cdot \frac{dI_p}{dt} - \frac{180R_o + 432R_c}{R_o(R_o + 2R_c)} \cdot \frac{dI_s}{dt} \tag{14}$$

$$V_s = \frac{180R_o + 432R_c}{R_o(R_o + 2R_c)} \cdot \frac{dI_p}{dt} - \frac{180R_o + 324R_c}{R_o(R_o + 2R_c)} \cdot \frac{dI_s}{dt} \tag{15}$$

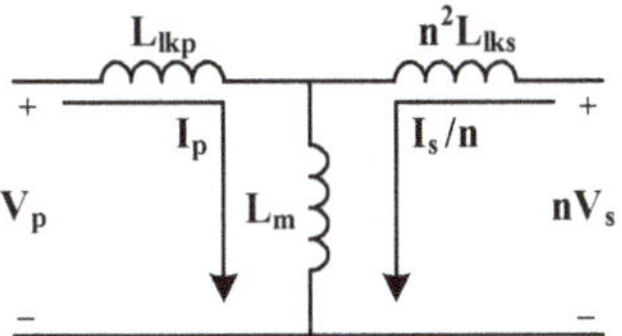

Figure 7. T model of the transformer.

By using Faraday's law of induction to assess the relationship between the voltage, reluctance, and number of turns of the transformer, the T model of the transformer (Figure 7) could be applied. In Figure 7, V_p refers to the primary-side voltage of the transformer, and V_s refers to the secondary-side voltage of the transformer; furthermore, L_{lkp} represents the primary-side leakage inductance, L_{lks} represents the secondary-side leakage inductance, and L_m represents the excitation inductance of the transformer. The T model enabled the equivalence of the voltage, current, and leakage inductance of the secondary side with that of the primary side. Equations (16) and (17) present the results.

$$V_p = \left(L_m + L_{lkp}\right)\frac{dI_p}{dt} - \left(\frac{L_m}{n}\right)\frac{dI_s}{dt} \tag{16}$$

$$V_s = -\left(\frac{L_m}{n}\right)\frac{dI_p}{dt} + \left(\frac{L_m}{n^2} + L_{lks}\right)\frac{dI_s}{dt} \tag{17}$$

Subsequently, Equations (14) and (16) could be compared with Equations (15) and (17). Equations (18)–(20) represent the relationship among the leakage inductance and excitation inductance of the primary and secondary sides of the transformer and the number of turns of the windings.

$$L_m + L_{lkp} - \frac{360R_o + 576R_c}{R_o(R_o + 2R_c)} \tag{18}$$

$$\frac{L_m}{n} = \frac{180R_o + 432R_c}{R_o(R_o + 2R_c)} \tag{19}$$

$$\frac{L_m}{n^2} + L_{lks} = \frac{180R_o + 324R_c}{R_o(R_o + 2R_c)} \tag{20}$$

The comparison could be used to establish models of equivalent leakage inductance of the primary and secondary sides of the transformer and the excitation inductance. Such models could be presented using Equations (21)–(23).

$$L_m = \frac{240R_o + 576R_c}{R_o(R_o + 2R_c)} \tag{21}$$

$$L_{lkp} = \frac{120R_o}{R_o(R_o + 2R_c)} \tag{22}$$

$$L_{lks} = \frac{45R_o}{R_o(R_o + 2R_c)} \tag{23}$$

According to the obtained results, a three-legged core could achieve adjustable leakage inductance if the shape of the outer legs easily enables winding. Applying the winding coil is difficult when the commonly used PQ and RM cores are employed because of their outer leg shape; thus, EI or EE cores were used for adjustable leakage inductance [21]. The simulation results of the current distribution of the PCB trace windings based on the shapes of the PQ and EI cores were obtained with Maxwell magnetic simulation software. The simulation results presented in Figure 8 yielded the winding for the square core, and the current distribution obtained using the square core was considerably more uneven than that obtained using the round core, leading to excessively dense current at the corner of the winding, hot spots, and higher losses. Therefore, the cylindrical core was selected as the transformer core in this study.

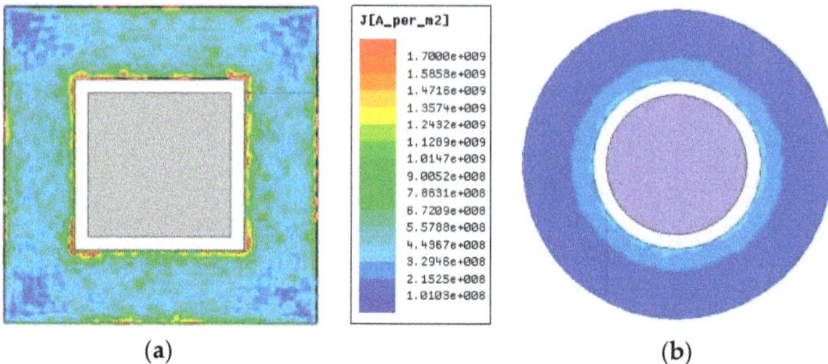

Figure 8. Relationship between the winding shapes of effective core cross sections and current distribution: (**a**) square-shaped cross section; (**b**) round cross section.

The design presented in Figure 9 was proposed to achieve adjustable inductance through winding on the two outer legs of the three-legged core and the obtainment of even current distributions on PCB windings. Producing adjustable leakage inductance required that windings be set only on the outer legs and not on the center leg. This method enabled the cross-sectional area of the center leg to be altered to a shape suitable for PCB winding, increasing the effectiveness of the winding space of the transformer. Therefore, this study proposed aligning two PQ cores (Figure 9a), where the connected part serves as the center leg of the core, and removal of one outer leg of the original core forms a new three-legged structure for the core. The design is as presented in Figure 9b.

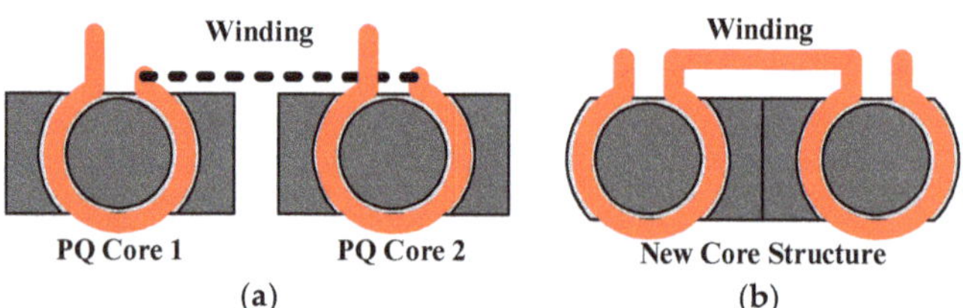

Figure 9. Diagram of the innovative core structure: (**a**) PQ core structure; (**b**) the innovative core structure.

Figure 10 shows a diagram of the core size design in this study. Figure 10a is the top view, and r represents the radius of the effective cross-sectional area, R denotes the radius of the core that can be wound, l is twice the horizontal distance from the core cylinder to the highest point of the center leg, m is the maximum width of the center leg, n is the length of the core, d is the thickness of the top and bottom pieces, and t represents the leg height of the core.

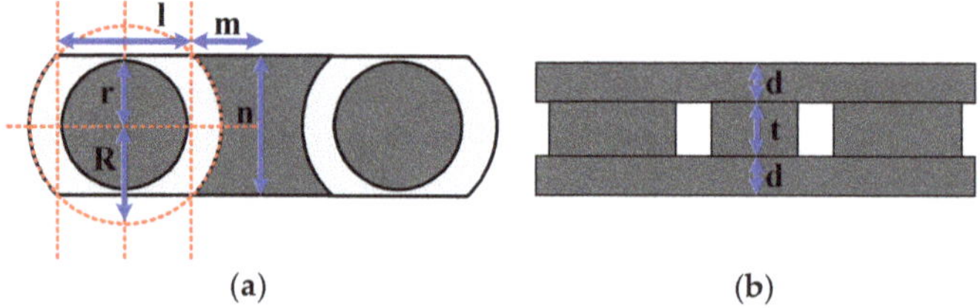

Figure 10. Diagram of the core size marking: (**a**) top and (**b**) side view.

The overall loss of the transformer can be divided into core loss and copper loss. Core loss is the product of core loss per unit volume and core volume, as shown in Equation (24). In Equation (25), *Pv* denotes the unit volume loss, and *Cm*, *x*, and *y* can be acquired from the core manufacturer's manual. Combining Equations (24) and (25) reveals that the core loss and volume *Vel* are related to the switching frequency *fs* and peak flux density B_{max}, with core loss being linearly proportional to *Vel* and exponentially proportional to *fs* and B_{max}. Next, B_{max} was expressed as Equation (26), where V_{in} denotes the input voltage, *Ae* represents the effective cross-sectional area of the core, and *Np* is the number of turns of the primary winding. In this study, the input voltage V_{in}, switching frequency *fs*, and effective cross-sectional area of the core are 400 V, 500 kHz, and 194 mm^2, respectively. The relationship between core loss and the number of primary-side turns is revealed through Equation (26), with a higher number of turns leading to a drop in peak flux density, reducing the core loss. However, from the perspective of copper loss, more turns resulted in greater loss. Thus, the optimal number of turns in this study was acquired by quantifying the core loss and copper loss.

$$Coreloss = Pcv \cdot Vel \tag{24}$$

$$Pv = Cm \cdot fs^x \cdot B_{max}^y \tag{25}$$

$$B_{max} = \frac{Vin}{4 \cdot Ae \cdot Np \cdot fs} \tag{26}$$

In this study, a full-bridge LLC resonant conversion topology was employed for the primary-side circuit architecture, whereas the secondary-side rectification architecture consisted of a full-bridge topology. The turns ratio of the transformer was calculated using Equation (27), and substituting the circuit specifications of this study (400 V input voltage and 300 V output voltage) yielded a transformer turns ratio of 4:3. However, the minimal number of turns on the primary side cannot be 1 because of the winding method of the adjustable leakage inductance, because the primary-side should be wound to the two outer legs, and because the primary side on both legs cannot be the same. Additionally, the winding ratio of the primary and secondary sides has to be an integer to prevent the need for a non-integer number of turns on the secondary side. Table 4 shows that the winding width of the core could be obtained by subtracting *r* from *R*, and the actual winding width was 5.5 mm.

$$n = \frac{Vin}{Vo} \tag{27}$$

Table 4. Parameter design of the asymmetric resonant tank.

Core Size	Value
Ae	194 mm^2
R	13.3 mm
r	7.8 mm
l	18.8 mm
m	7.29 mm
n	18.8 mm
d	6 mm
t	10.3 mm

The transformer turns ratio in this study was 4:3; thus, the number of turns at the primary and secondary sides could be 4:3, 8:6, 12:9, and so on. The core loss per unit volume (*Pv*) increased substantially when the peak flux density of the transformer exceeded 1000 G. Therefore, the peak flux density was designed to be less than 1000 G in the core turns design. Next, Equation (26) was rewritten as Equation (28), and B_{max} was set to 1000 G. According to Equation (28), the minimum number of winds on the primary side was 14.6. Practically, more windings mean more layers and greater costs.

Therefore, only the following five windings on the primary and secondary sides were considered: 16:12, 20:15, 24:18, 28:21, and 32:24.

$$Np > \frac{Vin}{4 \cdot Ae \cdot B_{max} \cdot fs} = 14.6 \tag{28}$$

The peak flux density and core loss for different numbers of turns were obtained using Equations (24) and (26), and the results are plotted in Figure 11. The relationship between core loss and number of turns was nonlinear. The difference in core loss between 16 and 20 turns on the primary-side winding was 4.8 W, whereas 28 and 32 turns on the primary-side winding yielded a core loss of approximately 0.8 W. This was the main reason for designing B_{max} to be less than 1000 G.

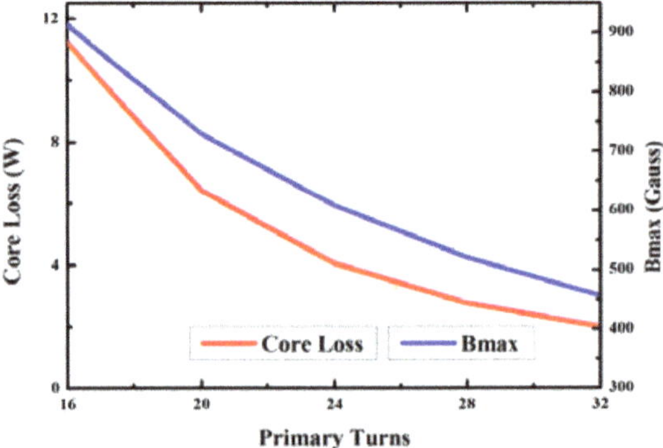

Figure 11. Peak flux density and core loss.

The high-frequency copper loss was analyzed next. Conventional transformers are mostly wound with litz wire; however, planar routing is more advantageous to utilize the window area of a transformer. Figure 12 compares the use of planar routing and four litz wires. Under the same condition of an occupied area of 4 mm^2, the cross-sectional area of the litz wires was 3.142 mm^2, whereas that of the planar routing was 4 mm^2 (Figure 12a,b, respectively); both yielded a 27% difference in DC resistance under the same length. Therefore, PCB routing was used as the transformer winding in this study to achieve high power density.

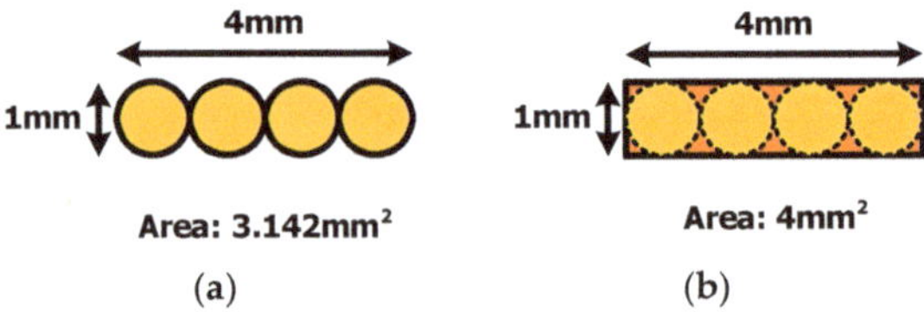

Figure 12. Winding usage in different transformers: (**a**) litz wire and (**b**) planar trace.

PCB routing was selected over litz wire under the same transformer window area because it resulted in lower DC resistance of the winding. However, a sharp rise in AC resistance loss was observed when the switching frequency was increased from the conventional 50 kHz–200 kHz to 500 kHz–1 MHz. Therefore, the AC resistance loss of the transformer winding was analyzed.

According to Dowell's equation, the skin effect and proximity effect can be expressed by Equations (29) and (30), respectively, where $\xi = h/\delta$, with h denoting the thickness of the copper wire in the transformer winding and δ representing the skin depth of the transformer winding. Additionally, m can be expressed using Equation (31), where e represents the number of winding layers. The five numbers of windings that were employed and the corresponding distributions of MMF are displayed in Figure 13.

Because the circuit architecture in this study was an LLC rather than bidirectional CLLC, only the leakage inductance on the primary side was adjusted. Therefore, the secondary side was wound around the two outer legs with the same or almost the same number of turns, and the primary and secondary sides were wound on an eight-layer PCB in an interleaved manner. A total of seven layers of PCB winding were used, and one layer was reserved as the routing that connected the two leg windings.

$$R_{ac\_skin} = \frac{\xi}{2} \cdot \frac{\sinh(\xi) + \sin(\xi)}{\cosh(\xi) - \cos(\xi)} \cdot R_{dc} \tag{29}$$

$$R_{ac\_proximity} = \frac{\xi}{2} \cdot (2m - 1)^2 \cdot \frac{\sinh(\xi) - \sin(\xi)}{\cosh(\xi) + \cos(\xi)} \cdot R_{dc} \tag{30}$$

$$m = \left| \frac{MMF(e)}{MMF(e) - MMF(e - 1)} \right| \tag{31}$$

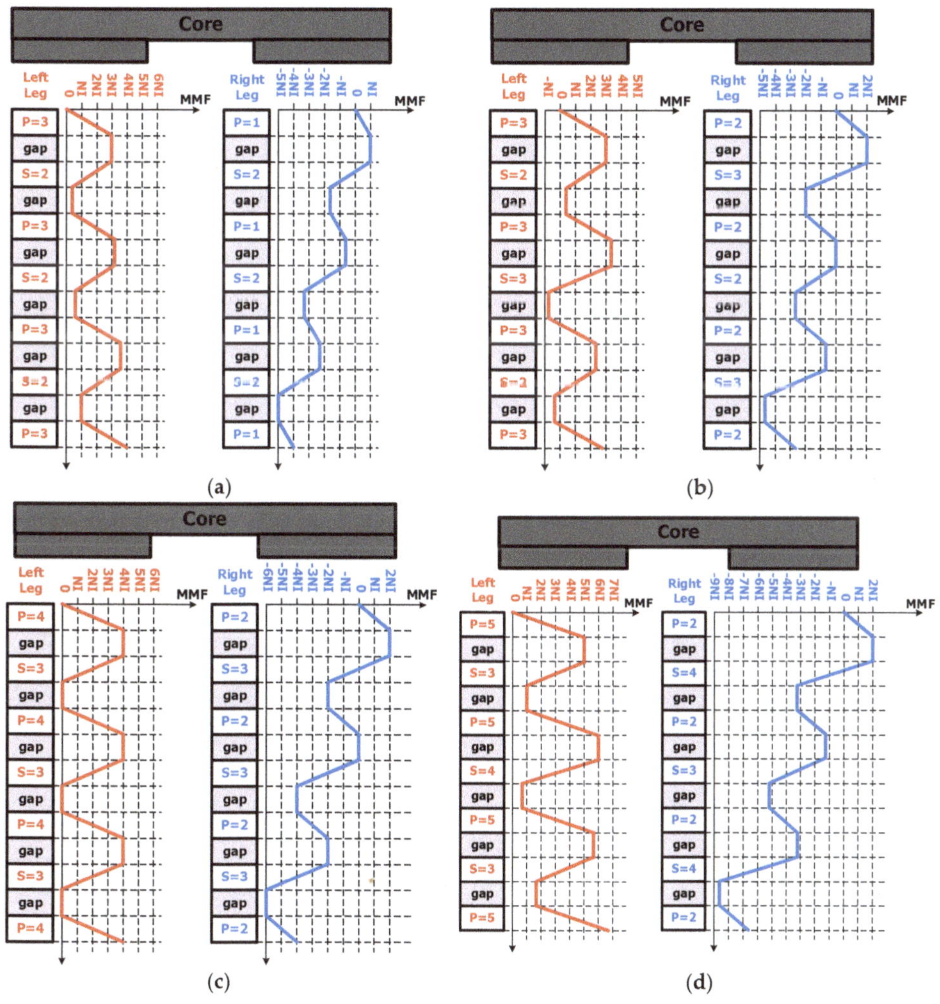

(a)　(b)

(c)　(d)

Figure 13. *Cont.*

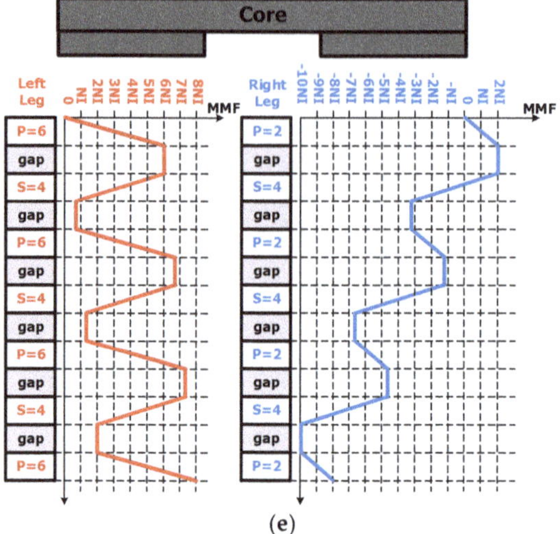

Figure 13. MMF distribution map of different numbers of windings: (**a**) $Np = 16$, $Ns = 12$; (**b**) $Np = 20$, $Ns = 15$; (**c**) $Np = 24$, $Ns = 18$; (**d**) $Np = 28$, $Ns = 21$; and (**e**) $Np = 32$, $Ns = 24$.

Figure 14 illustrates the copper loss and total loss of the transformer for different numbers of turns. The core loss could be obtained from Figure 11. More turns resulted in lower peak flux density of the core and thus lower core loss. However, more windings also led to greater copper loss, which could be divided into that associated with DC resistance and AC resistance. DC resistance was greater for more windings; however, the resistance that actually affects the high-frequency copper loss is AC resistance loss. Because an eight-layer board was used for transformer routing, more turns meant that each layer of the PCB board had to accommodate more windings, which narrowed the width of each winding. Moreover, according to Figure 13, more windings in each layer resulted in a higher MMF, causing substantially greater AC resistance under high frequency, which in turn resulted in greater copper loss. The various amounts of core loss and copper loss are presented in Table 5, and the results are plotted in Figure 14. Ultimately, the number of turns with the lowest total loss (24:18) was selected as the suitable number of turns of the transformer in this study. Figure 15 shows the peak flux density of the core, as simulated by ANSYS Maxwell software, to verify that the core operated normally.

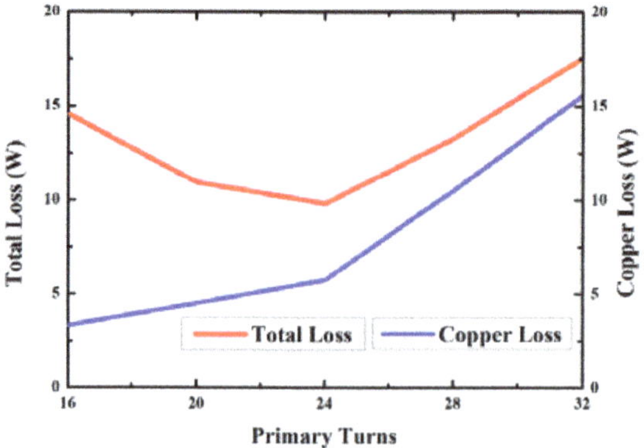

Figure 14. Copper loss and total loss.

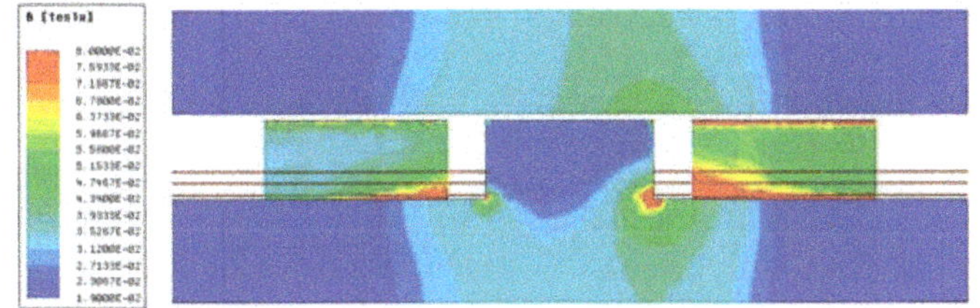

Figure 15. Flux density distribution by FEA 3D simulation.

Table 5. Parameter design of the asymmetric resonant tank.

Condition\Np:Ns	16:12	20:15	24:18	28:21	32:24
B_{max} (Gauss)	913	730	608	521	456
Core Loss (W)	11.23	6.43	4.07	2.77	1.99
Rdc (mΩ)	157	223	260	479	689
Rac (mΩ)	167	232	270	530	796
Copper Loss (W)	3.33	4.50	5.74	10.50	15.57
Total Loss (W)	14.56	10.93	9.81	13.27	17.55

4. Results

An LLC resonant converter with 1 kW output power and 500 kHz switching frequency was proposed in this study. Figure 16a,b shows the actual core size and circuit diagram, respectively. Table 6 details the components and design parameters used in this study.

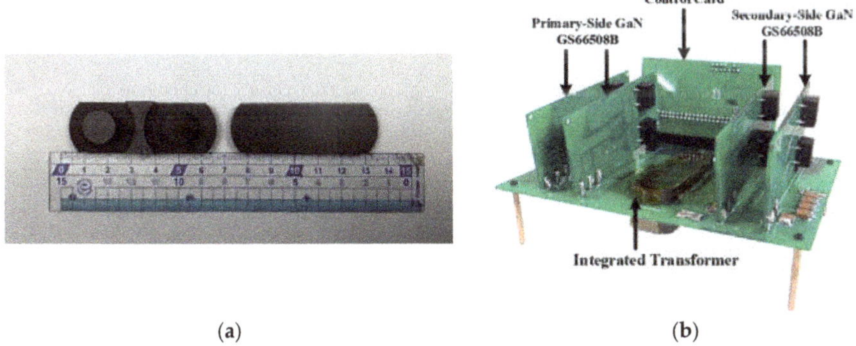

(a)   (b)

Figure 16. Experimental prototype of the (**a**) core structure and (**b**) circuit.

Table 6. Parameter design of the asymmetric resonant tank.

Component	Value
Resonant and Operating Frequency	500 kHz
Transformer Turns Ratio	24:18
Transformer Material	3F36
Primary-Side Switch Device	GS66508B
Secondary-Side Switch Device	GS66508B
Primary-Side Driver	Si8271GB
Secondary-Side Driver	Si8271GB
Resonant Capacitor	22.4 nF
Leakage Inductance	4.7 μH
Magnetizing Inductance	80.5 μH

Figure 17 plots the measured conversion efficiency of the converter. The highest efficiency of 97.03% was achieved at 80% load. Figure 18 displays the circuit test waveform at full load, which includes the waveforms of *Vds*, *Vgs*, and the resonant current I_{Lr} of the primary-side power switch operating properly. Figure 19a–c shows temperature distribution of the primary-side power components at full load, with maximum temperatures of 38.1 °C, 40.4 °C, and 78.6 °C, respectively. Because the specifications of this paper are high-voltage to high-voltage, according to the loss analysis in Figure 4, the loss can be mainly attributed to the transformer, especially, on the transformer windings. Because the primary or secondary switching elements have zero-voltage-switch turn-on, the temperature distribution of the overall converter will be concentrated on the windings of the transformer.

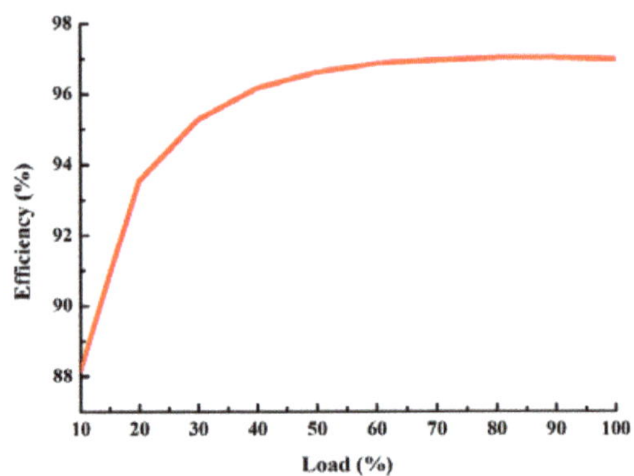

Figure 17. Measured power efficiency.

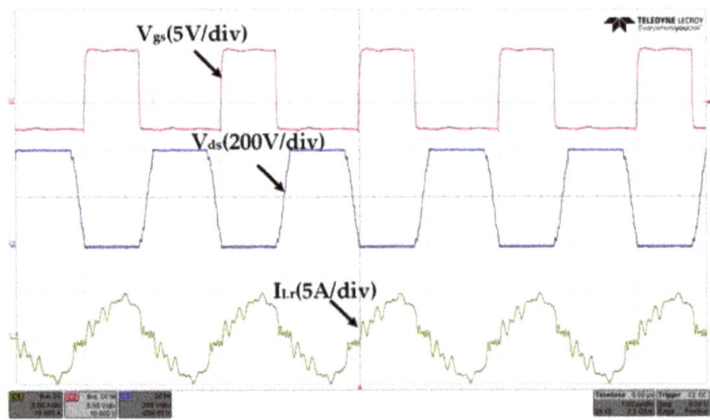

Figure 18. Experimental waveforms for the converter at full load.

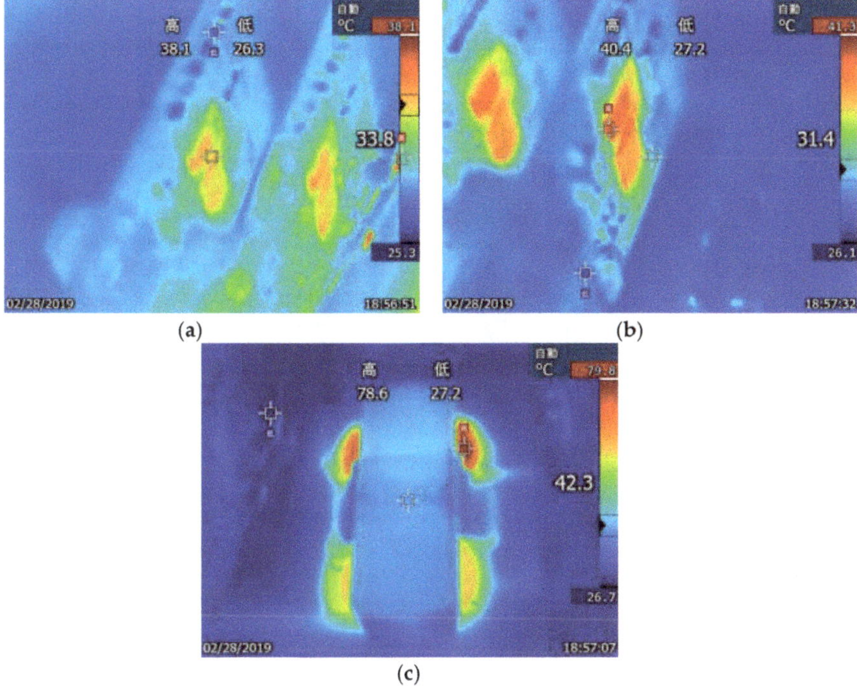

Figure 19. Temperature of the LLC converter at full load, (**a**) primary-side GaN devices, (**b**) secondary side GaN devices, and (**c**) core and winding of transformer.

5. Conclusions

This study determined the primary-side topology and secondary-side rectification topology suitable for an LLC resonant converter under a switching frequency, input voltage, and output voltage of 500 kHz, 400 V, and 300 V, respectively. After loss analysis and determining the rated voltage and current of the components, the full-bridge topology was selected for both the primary and secondary sides. This study also investigated the small transformer leakage inductance caused by an interleaved winding structure, which was employed to reduce the winding MMF of the transformer under high-frequency switching. The adjustable leakage inductance structure was thus adopted in this study, and the influence of core shape on current distribution was explored. Subsequently, the feasibility of the core was verified through simulation using ANSYS Maxwell software. Ultimately, an LLC resonant converter with 1 kW output power and 500 kHz switching frequency was designed and tested. The results verified a maximum conversion efficiency of 97.03%.

Author Contributions: Y.-C.L., C.C., K.-D.C. and Y.-L.S. conceived and designed the prototype and experiments; C.C., K.-D.C., Y.-L.S. and M.-C. Tsai performed the experiments; Y.-C.L., C.C., K.-D.C. analyzed the data; Y.-C.L. contributed equipment/materials/analysis tools; Y.-C.L., C.C., K.-D.C. wrote the paper.

Funding: This work was supported in part by the Ministry of Science and Technology, Taiwan under Grant 107-2218-E-008 -012- and 107-2221-E-197 -030 -.

References

1. Carli, R.; Dotoli, M.; Cianci, E. An optimization tool for energy efficiency of street lighting systems in smart cities. *IFAC-PapersOnLine* **2017**, *50*, 14460–14464. [CrossRef]
2. Pandharipande, A.; Caicedo, D. Smart indoor lighting systems with luminaire-based sensing: A review of lighting control approaches. *Energy Build.* **2015**, *104*, 369–377. [CrossRef]
3. Carli, R.; Dotoli, M.; Pellegrino, R. A decision-making tool for energy efficiency optimization of street lighting. *Comput. Oper.* **2018**, *96*, 223–235. [CrossRef]
4. Chen, X.; Huang, D.; Li, Q.; Lee, F.C. Multi-channel LED driver with CLL resonant converter. In Proceedings of the IEEE Energy Conversion Congress and Exposition (ECCE), Pittsburgh, PA, USA, 14–18 September 2014; pp. 3599–3606.
5. Shang, M.; Wang, H. A LLC type resonant converter based on PWM voltage quadrupler rectifier with wide output voltage. In Proceedings of the IEEE Applied Power Electronics Conference and Exposition (APEC), Tampa, FL, USA, 26–30 March 2017; pp. 1720–1726.
6. Liu, Y.-C.; Chen, K.-D.; Chen, C.; Syu, Y.-L.; Lin, G.-W.; Kim, K.A.; Chiu, H.-J. Quarter-Turn Transformer Design and Optimization for High Power Density 1-MHz LLC Resonant Converter. *IEEE Trans. Ind. Electron.* **2019**. [CrossRef]
7. Lu, J.L.; Hou, R.; Chen, D. Opportunities and design considerations of GaN HEMTs in ZVS applications. In Proceedings of the IEEE Applied Power Electronics Conference and Exposition (APEC), San Antonio, TX, USA, 4–8 March 2018; pp. 880–885.
8. Chen, R.; Brohlin, P.; Dapkus, D. Design and magnetics optimization of LLC resonant converter with GaN. In Proceedings of the IEEE Applied Power Electronics Conference and Exposition (APEC), Tampa, FL, USA, 26–30 March 2017; pp. 94–98.
9. Fei, C.; Gadelrab, R.; Li, Q.; Lee, F.C. High-Frequency Three-Phase Interleaved LLC Resonant Converter with GaN Devices and Integrated Planar Magnetics. *IEEE J. Emerg. Sel. Top. Power Electron.* **2019**, *7*, 653–663. [CrossRef]
10. He, P.; Khaligh, A. Comprehensive Analyses and Comparison of 1 kW Isolated DC–DC Converters for Bidirectional EV Charging Systems. *IEEE Trans. Transp. Electrif.* **2017**, *3*, 147–156. [CrossRef]
11. Dung, N.A.; Chiu, H.-J.; Lin, J.-Y.; Hsieh, Y.-C.; Chen, H.-T.; Zeng, B.-X.; Nguyen, D.A. Novel Modulation of Isolated Bidirectional DC–DC Converter for Energy Storage Systems. *IEEE Trans. Power Electron.* **2019**, *34*, 1266–1275. [CrossRef]
12. Huang, D.; Ji, S.; Lee, F.C. LLC resonant converter with matrix transformer. *IEEE Trans. Power Electron.* **2014**, *29*, 4339–4347. [CrossRef]
13. Mu, M.; Lee, F. Design and Optimization of a 380V-12V High-Frequency, High-Current LLC Converter with GaN Devices and Planar Matrix Transformers. *IEEE J. Emerg. Sel. Top. Power Electron.* **2016**, *4*, 1. [CrossRef]
14. Fei, C.; Lee, F.C.; Li, Q. High-Efficiency High-Power-Density LLC Converter with an Integrated Planar Matrix Transformer for High-Output Current Applications. *IEEE Trans. Ind. Electron.* **2017**, *64*, 9072–9082. [CrossRef]
15. Dowell, P. Effects of eddy currents in transformer windings. *Proc. Inst. Electr. Eng.* **1966**, *113*, 1387. [CrossRef]
16. Dixon, L.H. Eddy Current Losses in Transformer Winding and Circuit Wiring, Unitrode/TI Magnetics Design Handbook. 2000.
17. Ouyang, Z.; Hurley, W.G.; Andersen, M.A.E. Improved Analysis and Modelling of Leakage Inductance for Planar Transformers. *IEEE J. Emerg. Sel. Top. Power Electron.* **2018**. [CrossRef]
18. Liu, Y.-C.; Chen, C.; Lin, S.-Y.; Xiao, C.-Y.; Kim, K.A.; Hsieh, Y.-C.; Chiù, H.-J. Integrated magnetics design for a full-bridge phase-shifted converter. In Proceedings of the IEEE Applied Power Electronics Conference and Exposition, San Antonio, TX, USA, 4–8 March 2018; pp. 2110–2116.
19. Abu Bakar, M.; Bertilsson, K. An improved modelling and construction of power transformer for controlled leakage inductance. In Proceedings of the IEEE International Conference on Environment and Electrical Engineering (EEEIC), Florence, Italy, 7–10 June 2016; pp. 1–5.
20. Liu, G.; Li, D.; Zhang, J.Q.; Hu, B.; Jia, M.L. Bidirectional CLLC resonant DC-DC converter with integrated magnetic for OBCM application. In Proceedings of the IEEE International Conference on Industrial Technology (ICIT), Seville, Spain, 17–19 March 2015; pp. 946–951.

21. Li, B.; Li, Q.; Lee, F.C. A novel PCB winding transformer with controllable leakage integration for a 6.6kW 500kHz high efficiency high density bi-directional on-board charger. In Proceedings of the 2017 IEEE Applied Power Electronics Conference and Exposition (APEC), Tampa, FL, USA, 26–30 March 2017; pp. 2917–2924.

22. Li, B.; Li, Q.; Lee, F. High Frequency PCB Winding Transformer with Integrated Inductors for a Bi-directional Resonant Converter. *IEEE Trans. Power Electron.* **2019**, *34*, 6123–6135. [CrossRef]

23. Alonso, J.M.; Vina, J.; Vaquero, D.G.; Martinez, G.; Osorio, R. Analysis and Design of the Integrated Double Buck–Boost Converter as a High-Power-Factor Driver for Power-LED Lamps. *IEEE Trans. Ind. Electron.* **2012**, *59*, 1689–1697. [CrossRef]

24. Chen, W.; Hui, S.Y.R. Elimination of an Electrolytic Capacitor in AC/DC Light-Emitting Diode (LED) Driver with High Input Power Factor and Constant Output Current. *IEEE Trans. Power Electron.* **2012**, *27*, 1598–1607. [CrossRef]

25. Qiu, Y.; Wang, L.; Wang, H.; Liu, Y.-F.; Sen, P.C. Bipolar Ripple Cancellation Method to Achieve Single-Stage Electrolytic-Capacitor-Less High-Power LED Driver. *IEEE J. Emerg. Sel. Top. Power Electron.* **2015**, *3*, 698–713.

26. Ouyang, Z.; Thomsen, O.C.; Andersen, M.A.E. Optimal Design and Tradeoff Analysis of Planar Transformer in High-Power DC–DC Converters. *IEEE Trans. Ind. Electron.* **2012**, *59*, 2800–2810. [CrossRef]

27. Yang, B.; Lee, F.C.; Zhang, A.J.; Huang, G. LLC resonant converter for front end DC/DC conversion. In Proceedings of the IEEE Applied Power Electronics Conference and Exposition (Cat. No.02CH37335), Dallas, TX, USA, 10–14 March 2002; Volume 2, pp. 1108–1112.

Article

ZVS Auxiliary Circuit for a 10 kW Unregulated LLC Full-Bridge Operating at Resonant Frequency for Aircraft Application

Yann E. Bouvier [1,*] , Diego Serrano [1], Uroš Borović [1], Gonzalo Moreno [2], Miroslav Vasić [1], Jesús A. Oliver [1], Pedro Alou [1], José A. Cobos [1] and Jorge Carmena [2]

[1] Centro de Electrónica Industrial, Universidad Politécnica de Madrid, Madrid 28006, Spain; diego.serrano@upm.es (D.S.); uros.borovic@upm.es (U.B.); Miroslav.vasic@upm.es (M.V.); jesusangel.oliver@upm.es (J.A.O.); pedro.alou@upm.es (P.A.); ja.cobos@upm.es (J.A.C.)

[2] Indra Sistemas, S.A., Indra Sistemas, Calle de Joaquín Rodrigo, 11, 28300 Aranjuez, Madrid, Spain; gmorenoh@indra.es (G.M.); jcarmena@indra.es (J.C.)

* Correspondence: yann.bouvier@upm.es

Received: 12 April 2019; Accepted: 13 May 2019; Published: 15 May 2019

Abstract: In modern aircraft designs, following the More Electrical Aircraft (MEA) philosophy, there is a growing need for new high-power converters. In this context, innovative solutions to provide high efficiency and power density are required. This paper proposes an unregulated LLC full-bridge operating at resonant frequency to obtain a constant gain at all loads. The first harmonic approximation (FHA) model is not accurate enough to estimate the voltage gain in converters with high parasitic resistance. A modified FHA model is proposed for voltage gain analysis, and time-based models are used to calculate the instantaneous current required for the ZVS transition analysis. A method using charge instead of current is proposed and used for this ZVS analysis. Using this method, an auxiliary circuit is proposed to achieve complete ZVS within the whole load range, avoiding a gapped transformer design and increasing the efficiency and power density. A 28 Vdc output voltage prototype, with 10 kW peak output power, has been developed to validate the theoretical analysis and the proposed auxiliary circuit. The maximum efficiency (96.3%) is achieved at the nominal power of 5 kW.

Keywords: aircraft power conversion; LLC resonant converters; high efficiency; ZVS auxiliary circuit

---

1. Introduction

In the field of the More Electrical Aircraft (MEA) philosophy, there is a tendency to substitute mechanical, hydraulic and pneumatic systems with their electrical equivalents in order to increase efficiency and reduce cost and fuel consumption [1–3]. As a result, the electrical power demand of modern aircraft has increased and new power conversion solutions are needed.

Aircraft generators typically supply electrical power in AC at variable frequency (360 Hz/800 Hz; see Figure 1); however, aircraft loads require 28 Vdc [4]. Traditional aircraft rectifiers are based on passive solutions, with low-frequency transformers and diode rectifications. This approach is very robust, but the power density is low and offers limited regulation capabilities. Following the MEA philosophy, active rectifier solutions are being developed. These active solutions can be more efficient, with an optimized volume and weight, and can also reduce the harmonic content in the AC grid, which increases the lifecycle of the generators. Active aircraft rectifier systems are typically composed of three stages, as shown in Figure 1: an EMI filter, AC/DC rectifier and isolated DC/DC converter.

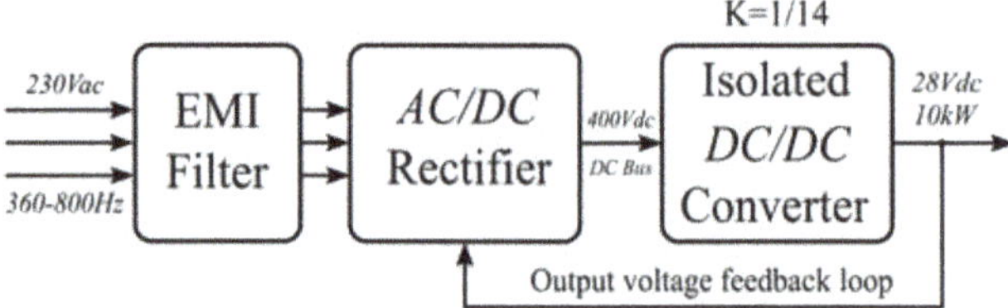

Figure 1. Aircraft active isolated rectifier architecture.

This article will focus on the DC/DC stage following the specifications in Table 1; a previous publication on the design can be found in [5]. Additional information on the rectifier stage can be found in [6]. The output voltage is fed and controlled by the rectifier, as depicted in Figure 1. Therefore, the DC/DC converter needs a constant gain (K = 1/14) for all loads.

Table 1. Isolated DC/DC converter specification.

Input Voltage	Output Voltage	Output Power	Output Current
400 V	28 V	10 kW	360 A

For a high-power isolated converter, full-bridge based topologies are known to be the most adequate solution in terms of volume and losses [7]. High-frequency operation is desirable to decrease the magnetic component volume and weight; however, switching losses in the devices will increase. That is why soft-switched topologies offer a good trade-off between efficiency and volume, such as with LLC converters, which are widely used in different applications [8,9].

The topology selected for this application is the LLC full-bridge operating at resonant frequency [10–12], at which the maximum efficiency is reached. However, this operation point is at the crossroad between zero voltage switching (ZVS) and zero current switching (ZCS) for primary devices. ZCS and ZVS are mutually exclusive soft switching techniques and, in high-current converters, ZCS achieves a higher loss reduction than ZVS. However, aircraft noise standards are restrictive. Therefore, ZVS operation must be achieved in high-voltage converters to avoid voltage spikes and ensure robust operation and low emitted noise. The ZVS capabilities within the whole load range have been analyzed extensively in the literature [13,14]. However, since operation at resonant frequency is between the ZCS and ZVS regions, time-based (TB) simulations are needed to estimate the range of ZVS depending on different LLC design parameters. Additionally, in LLC converters, the current is not constant along the ZVS transition and must be analyzed in terms of charge rather than current using time-based models. This is not considered in previous publications.

To achieve complete ZVS within the whole load range in unregulated LLC converters, the design leads to high circulating currents and a gapped transformer. However, auxiliary circuits can be used to increase the ZVS range in full-bridge converters; some are active [15] and others are passive [16–18], and they consist typically of two inductive circuits connected to the middle point of each leg. This work proposes a new configuration for the ZVS auxiliary circuit, specific to LLC full-bridge topology, with a single inductor connected between both middle points, reducing the circuit to a single inductor. With this circuit, ZVS can be achieved for the whole load range with an optimized transformer design without gaps and with LLC parameters optimized to reduce circulating currents.

First harmonic approximation (FHA) is an analytical method to estimate the voltage gain in resonant converters [19]. There are modifications of the FHA method in the literature, to account for higher harmonics [20,21]. However, these methods do not consider load-dependent behavior. In regulated LLC converters, the control loop can compensate for the error in gain. In unregulated LLC converters, an accurate estimation of the voltage gain mismatch is required. The unregulated LLC converter in this work has an external control loop to compensate the voltage gain error. However, an accurate voltage gain is required to estimate the components maximum voltage for the DC/DC and AC/DC stages from Figure 1.

Two modified FHA methods for voltage gain estimation are proposed in this paper and compared with their time-based equivalents. The first method includes a series resistance to account for the load-dependent voltage drop and the second method includes the distributed impedance model of the transformer.

Using the ZVS analysis and the modified FHA method, this work proposes an unregulated LLC full-bridge operating at resonant frequency with an auxiliary circuit to achieve complete ZVS within the whole load range. A 10 kW prototype is constructed to validate the topology and the models.

2. LLC Full-Bridge Topology: Operation Principle and Accurate Modeling

In this article, the considered topology is an unregulated LLC full-bridge with full-bridge rectifier, as shown in Figure 2. The duty cycle is constant and close to 50%, allowing a constant dead-time to achieve the ZVS transitions.

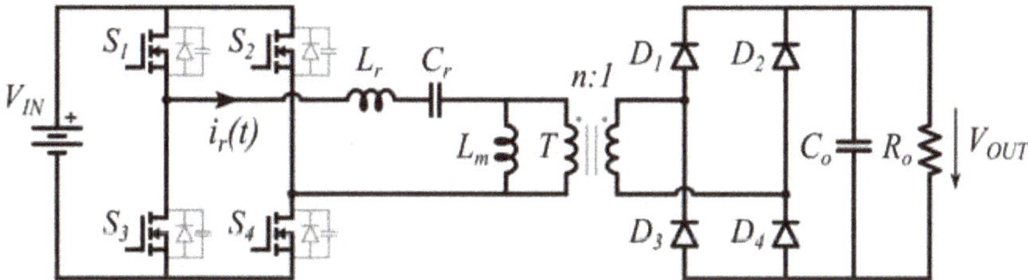

Figure 2. Isolated LLC full-bridge with full-bridge diode rectifier.

Normalization is performed with the parameters defined in Table 2.

Table 2. LLC converter normalized parameters.

Normalized frequency	Inductance ratio
$f_n = \frac{f_s}{f_r} = 2\pi f_s \sqrt{L_r C_r}$	$m = \frac{L_m + L_r}{L_r}$
Load quality factor	**Normalized voltage gain**
$Q = \frac{1}{R_O^*} \sqrt{\frac{L_r}{C_r}}$	$M = \frac{n V_{OUT}}{V_{IN}}$

In Table 2, f_s is the switching frequency and f_r is the resonant frequency of the series tank, and R_O^* is the reflected load AC-equivalent resistance for the output full-bridge rectifier. The rest of the parameters are defined in Figure 2.

The analytical equation for the gain is derived from the equivalent circuit shown in Figure 3a using the conventional first harmonic approximation (FHA) [19]:

$$M = \frac{V_2}{V_1} = \frac{n \cdot V_{OUT}}{V_{IN}} = \frac{1}{\sqrt{\left(1 + \frac{1}{m-1}\left(1 - \frac{1}{f_n^2}\right)\right)^2 + Q^2\left(f_n - \frac{1}{f_n}\right)^2}} \tag{1}$$

Figure 3b shows the gain (M) at different quality factors (Q) and inductance ratios (m) within a range of normalized frequencies ($0.5 < f_n < 2$).

Three different modes can be defined for the operation of the LLC converter: below resonance ($f_n < 1$), at resonance ($f_n = 1$) and above resonance ($f_n > 1$). The current shapes of each mode are shown in Figure 4 at two different inductance ratios (m).

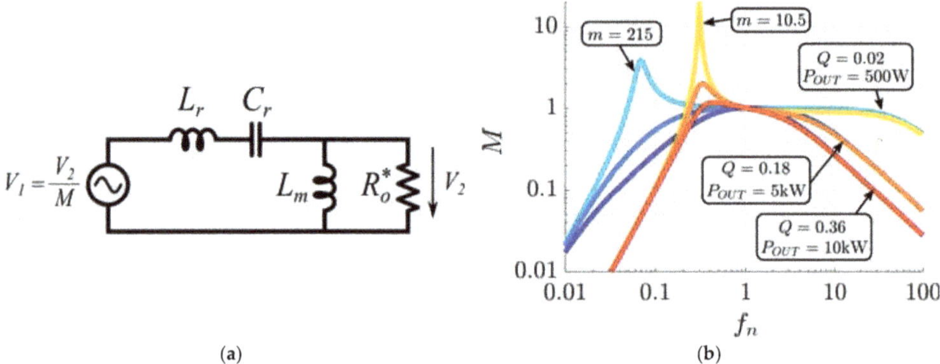

Figure 3. (**a**) LLC equivalent circuit (simplified first harmonic approximation (FHA)), (**b**) LLC voltage gain (*M*) at different inductance ratios (*m*) and quality factors (*Q*).

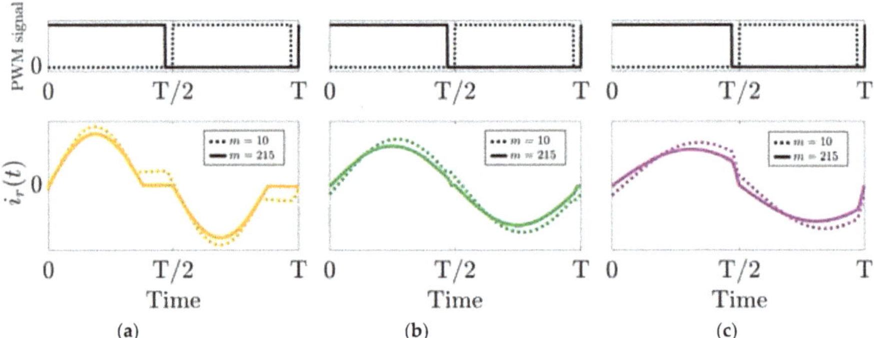

Figure 4. LLC converter modes: (**a**) below resonance; (**b**) at resonant; (**c**) above resonance.

2.1. Unregulated LLC Optimum Design and Power Loss Model

In regulated LLC converters, the minimum and maximum operating frequencies are selected to achieve the desired voltage gain. However, in unregulated LLC converters, to achieve a constant gain at all loads, the frequency is kept constant. A constant gain is only possible when operating at resonant frequency. Additionally, operation at resonant frequency achieves the best trade-off of power losses, as depicted in Figure 5. The model used for the MOSFET power loss can be found in the Infineon application note [22]. The conduction losses for a full-bridge configuration of MOSFET are calculated with the following equation:

$$P_C = \frac{4R_{DSon}(100\,^\circ C)}{N_{MOS}} I^2_{RMS\_M} \tag{2}$$

where $R_{DSon}(100\,^\circ C)$ is the drain-source resistance of the MOSFET with a junction temperature of $100\,^\circ C$ as a worst case, N_{MOS} is the number of devices in parallel for a switch position, and I_{RMS_M} is the switch total RMS current:

$$P_s = f_s N_{MOS} V_{DSMAX}\left(I_{DON}\frac{t_{ri}+t_{fu}}{2} + Q_{rr} + I_{DOFF}\frac{t_{ru}+t_{fi}}{2}\right), \tag{3}$$

where V_{DSMAX} is V_{IN} for primary-side and V_{OUT} for secondary-side devices, I_{DON} and I_{DOFF} are the turn-on and turn-off drain currents, $t_{ri}, t_{fu}, t_{ru}, t_{fi}$ are the rise and fall times of current and voltage, respectively, and finally Q_{rr} is the reverse recovery charge.

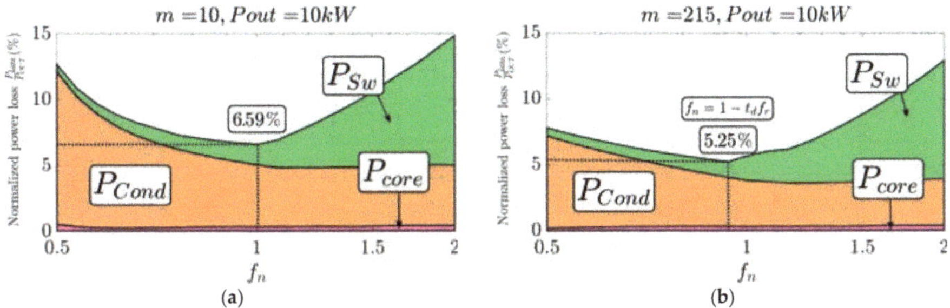

Figure 5. LLC converter total normalized power losses (P_{Loss}/P_{OUT}) at 10 kW output power. Including primary and secondary transistors and transformer. (**a**) for $m = 10$; (**b**) for $m = 215$.

To calculate the winding losses, the following equation is used:

$$P_{winding} = R_{AC}(100 \text{ kHz})\, I^2_{RMS_T} \tag{4}$$

where I_{RMS_T} is the transformer RMS current and $R_{AC}(100 \text{ kHz})$ is the AC resistance of the winding estimated using FEA analysis, using Pemag–Maxwell.

Finally, the core losses are estimated using the improved generalized Steinmetz equation (*iGSE*) [23]:

$$P_{core} = \frac{1}{T} \int_0^T k_i \left| \frac{dB}{dt} \right|^\alpha (\Delta B)^{\beta - \alpha} dt \tag{5}$$

where B is the flux in the core, α, β are the Steinmetz coefficients, and k_i is the improved Steinmetz coefficient. The components used can be found in Table 6 in Section 5. To summarize the total losses from breakdown in Figures 5 and 6, the losses are merged into three main categories of conduction losses,

$$P_{Cond} = P_{C1} + P_{C2} + P_{winding} + P_{Parasitics,} \tag{6}$$

consisting of the sum of the transistor conduction losses from primary and secondary sources, the winding losses of the transformer and the conduction losses of the parasitic resistances across the converter. Then, the switching losses consist of the sum of primary and secondary transistor switching losses, using Equation (3), and the core losses using Equation (5).

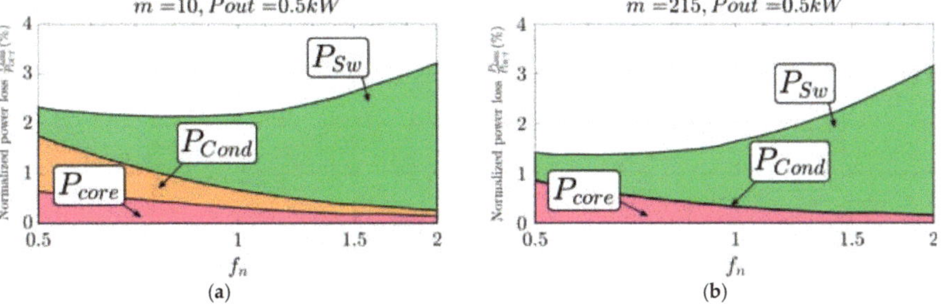

Figure 6. LLC converter total normalized power losses (P_{Loss}/P_{OUT}) at 500 W output power, including primary and secondary transistors and a transformer; (**a**) for $m = 10$; (**b**) for $m = 215$.

In Figure 5, for high inductance ratios (m), minimum losses are not reached at $f_n = 1$ but at $f_n = 1 - t_d \cdot f_r$, where complete ZCS is achieved. When comparing the behavior between high and low inductance ratios with the LLC converter, we can see in Figure 5a,b that $m = 215$ has lower power losses. This is caused by the high circulating currents and high turn-off currents of the $m = 10$ case.

These high circulating currents can be noticed at low load, as depicted in Figure 6, where the conduction losses are negligible for $m = 215$ but are considerable for $m = 10$.

The voltage gain is $M = 1$, when neglecting the load-dependent voltage drop and the effects of undesired parasitic elements. These effects are usually neglected in regulated LLC converters, because the feedback loop adjusts the voltage gain to the desired value. However, in unregulated LLC converters, the frequency is constant and the gain varies as a function of the load. In the system depicted in Figure 1, the DC/DC stage input voltage will be adjusted by the AC/DC stage. However, the input voltage range is required for the design of the AC/DC stage.

Because the converter is unregulated, the resonant frequency of the tank changes with the component value drift. If the switching frequency is not changed to match the new resonant frequency, the converter will not operate at the optimum efficiency. This issue can be solved by measuring the input and output voltage and implementing an algorithm to search for the maximum gain, which will also be the point of maximum efficiency. This is not a control loop, and it can be performed only once during the startup process. As explained before, the voltage gain varies with the output load and the conventional FHA does not consider this load influence and other parasitics. Therefore, more accurate modified models are needed to design unregulated LLC converters.

The next sections describe two FHA models with modified circuits with different parasitic elements added.

2.2. FHA and TB Models Considering the Series Resistance

As discussed previously, the effect of the load-dependent voltage drop cannot be neglected to estimate the gain. To consider this effect, series resistance is added to the equivalent circuit in in Figure 3a, as shown in Figure 7. The modified FHA model in Figure 7a is compared to the time-based (TB) model of Figure 7b. The addition of the resistance dampens the series resonance of the LLC tank, so higher harmonics are introduced in the current and non-linearities are no longer negligible. Therefore, the conventional FHA circuit is not accurate when the harmonic content is high.

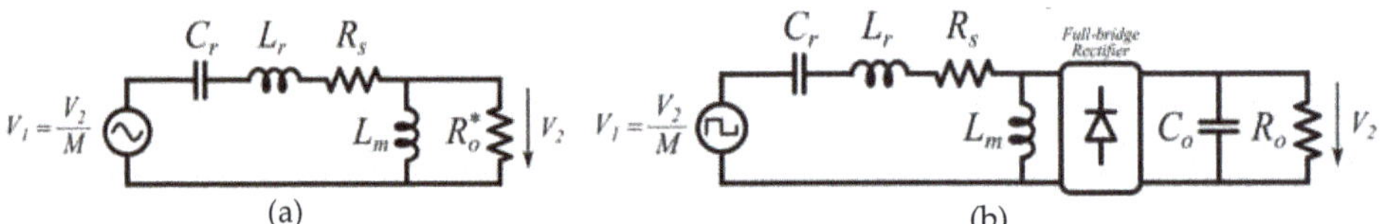

Figure 7. (a) Modified FHA LLC equivalent circuit with series resistance; (b) time-based (TB) LLC circuit with series resistance.

The analytical equation for the gain is derived from the modified FHA circuit, as illustrated in Figure 7a:

$$M = \frac{1}{\sqrt{\left(1 + \frac{1}{(m-1)}\left(1 - \frac{1}{f_n^2}\right) - \frac{Q}{Q_s}\right)^2 + Q^2 \cdot \left(f_n - \frac{1}{f_n} - \frac{1}{f_n^2} \cdot \left(\frac{1}{(m-1)QQ_s}\right)\right)^2}} \tag{7}$$

This equation is the same as Equation (1) with additional terms in bold. These terms contain a new parameter, the quality factor of the series resonant tank, which is defined as follows:

$$Q_s = \frac{1}{R_s}\sqrt{\frac{L_r}{C_r}} \tag{8}$$

This parameter includes the series resistance R_s, as depicted in Figure 7. It consists of all the resistances in series between the primary and secondary voltage bridges, including the resistances of the primary and secondary switches.

A comparison between the modified FHA and the TB models is depicted in Figure 8. The effect of the voltage drop is visible, as the gain at different power loads does not converge to one point at the resonant frequency. There is a good match between the modified FHA and TB models within the range of $0.75 < f_n < 1$. Outside this range, the harmonic content of the resonant current is high enough so that the FHA approach is not valid, especially at high power. Therefore, outside this range, the TB models should be used over the modified FHA for higher accuracy.

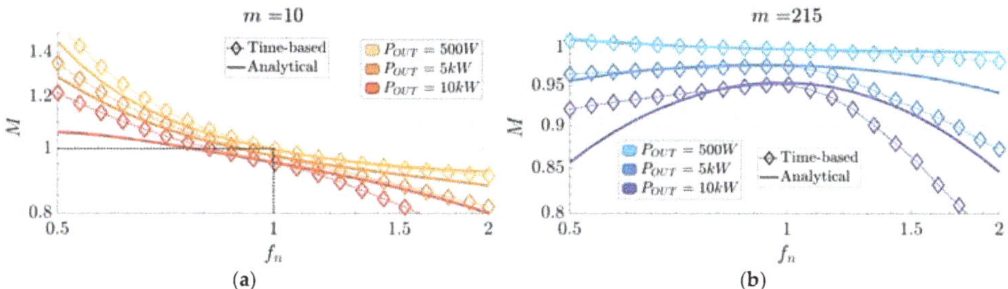

Figure 8. Comparison of the gain (M) for the modified FHA and TB models (with simulation) with series resistance at different output powers (**a**) for $m = 10$; (**b**) for $m = 215$.

2.3. FHA and TB Models Considering Distributed Impedance Model

The use of magnetic integration is widespread in LLC converters in the literature [24–26]. With magnetic integration, the two inductors of the LLC converter can be integrated in the transformer as a single magnetic component, where the magnetizing inductance operates as the parallel inductor L_m and the leakage inductance as the series resonant inductor L_r. However, the equivalent circuit of the transformer is more accurate with distributed leakage inductance if the inductance ratio (m) is low, as depicted in Figure 9. Considering external connections, stray inductances and resistances from the PCB and other parasitics, the equivalent model of the distributed LLC can be summarized as the circuit depicted in Figures 10 and 11.

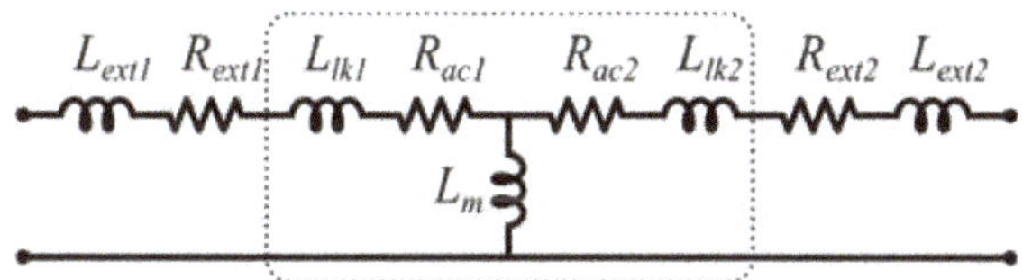

Figure 9. Equivalent circuit for transformer and external parasitics.

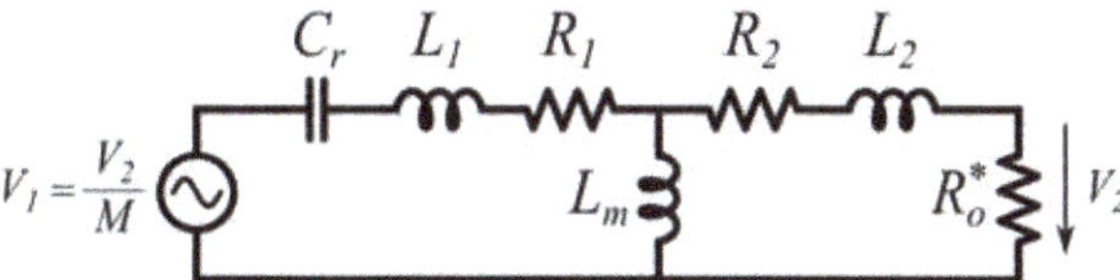

Figure 10. Modified FHA LLC equivalent circuit with distributed inductances and resistances.

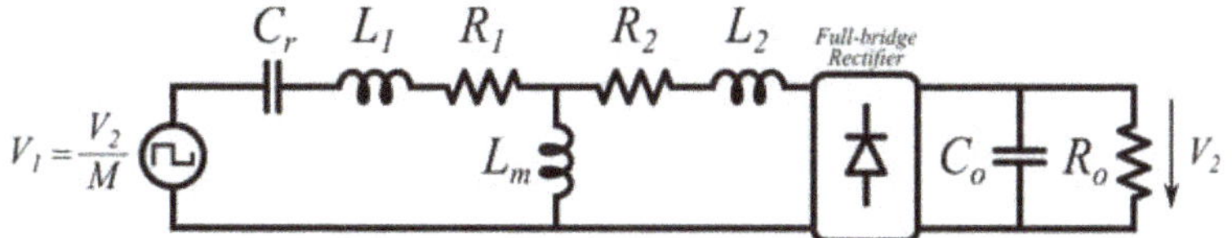

Figure 11. TB LLC circuit with distributed inductances and resistances.

The values of the distributed inductors and resistances are as follows:

$$(a)\ R_1 = \rho R_s \quad (b)\ R_2 = (1-\rho)R_s \quad (c)\ L_1 = \lambda L_r \quad (d)\ L_2 = (1-\lambda)L_r \tag{9}$$

where ρ and λ are expressed as numbers between 0 and 1 that establish the distribution between the primary and total series resistance and inductance. The analytical expression for the gain M can be calculated as follows:

$$M = \frac{1}{\sqrt{\left(1 + \frac{1}{(m-1)}\left(1 - \frac{(1+A_1)}{f_n^2}\right) - \frac{Q}{Q_s} + A_2\right)^2 + Q^2\left((1+B_1)f_n - \frac{(1+B_2)}{f_n} - \frac{1}{f_n^2}\cdot\left(\frac{1}{(m-1)QQ_s}\right)\right)^2}} \tag{10}$$

where the coefficients A_1, A_2, B_1 and B_2 are as follows:

$$(a)\ A_1 = (1-\rho)\frac{Q}{Q_s} \qquad (b)\ A_2 = \frac{1}{m-1}[\rho(1-\lambda) + (1-\rho)\lambda]\frac{Q}{Q_s} - (1-\lambda)$$
$$(c)\ B_1 = \frac{1-\lambda}{m-1} \qquad (d)\ B_2 = \frac{1}{m-1}\left[(1-\lambda) + \frac{\rho(1-\rho)}{Q_s^2} - \frac{(1-\rho)}{QQ_s}\right] \tag{11}$$

It can be observed that when $\rho = 1$ and $\lambda = 1$, the equivalent circuit becomes that shown in Figure 7, the coefficients A_1, A_2, B_1 and B_2 are equal to 0, and Equation (9) yields the same results as Equation (6). All of the coefficients in (10) depend on $1/(m-1)$, except for A_1, because it is already factorized in Equation (9). For this reason, the distributed model only affects the value of the gain for low values of m. This can be observed in Figure 12a,b, where the gain for $m = 215$ is the same for both models, while for $m = 10$, the gain increases by 10% at resonant frequency.

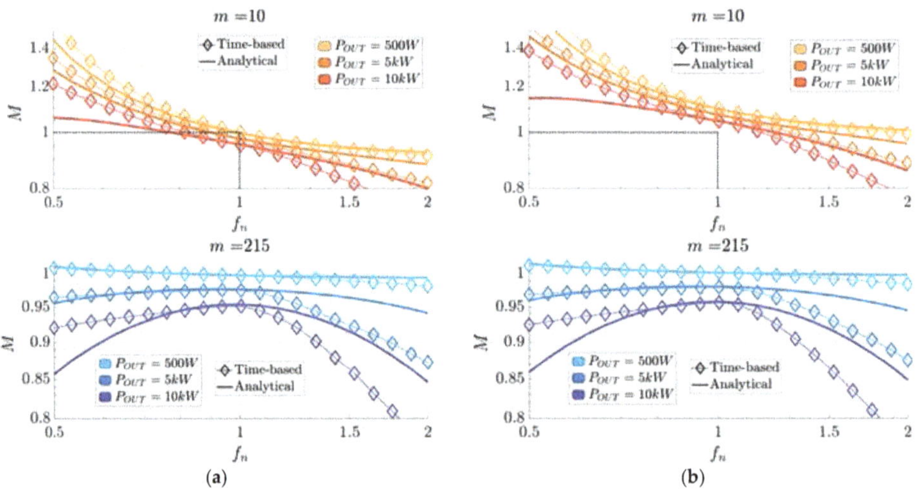

Figure 12. Comparison of the gain (M) for the series resistance model (Figure 5a) and distributed model (Figure 5b) with the FHA model and time-based simulations: (**a**) series resistance model gain at different powers and inductance ratios (Figure 6); (**b**) distributed model gain at different powers and inductance ratios.

In conclusion, for high values of m, the distributed model is not necessary and the series resistance model can be used as it gives a similar result. However, for low values of m, the distributed impedance model should be used because it is more accurate.

2.4. Time-Based Models for Current Analysis

The FHA models are based on the first harmonic; as such, the currents are purely sinusoidal. In the analysis of the ZVS transition, the instantaneous current is required and the FHA models are not accurate enough. Time-based models (TB) are then required for an accurate estimation of the instantaneous currents.

Two different TB models can be used, as explained in the previous two subsections: the series resistance model and the distributed impedance model, presented in Figure 5. (a) Modified FHA LLC equivalent circuit with series resistance; (b) time-based (TB) LLC circuit with series resistance. Figures 5b and 9, respectively. The distributed impedance model is closer to reality and gives more accurate results for low frequencies and low inductance ratios, as depicted in the next section in Figure 11.

In the next section, the ZVS transition will be analyzed in detail. However, it is necessary to know which of the two models is required for accurate modeling of the current during this transition.

3. ZVS Analysis for the Unregulated LLC Converter

The ZVS and ZCS regions of the LLC converter are depicted in Figure 13. ZVS and ZCS are mutually exclusive, because a turn-off current is needed to achieve a complete ZVS.

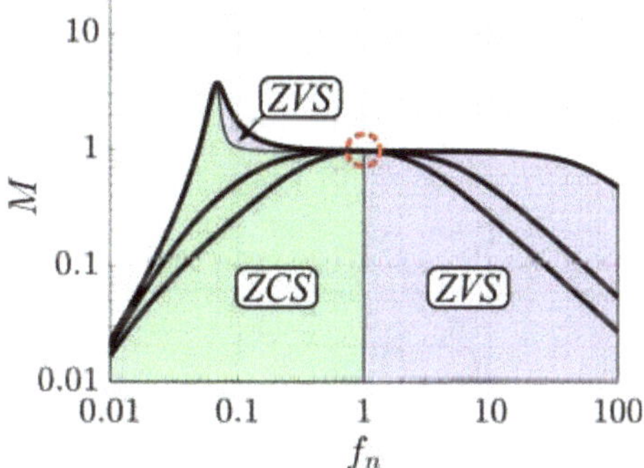

Figure 13. Zero voltage switching (ZVS) regions of the LLC converter.

The selected topology operates at the resonant frequency where the ZVS regions and the ZCS region meet. In this region, if the turn-off current is small enough, ZCS is not achieved and the ZVS transition will be incomplete (iZVS).

In power converters, ZVS is generally achieved using an inductive current to ensure the charge/discharge transition of the transistor output capacitance (C_{oss}). In the literature [27], this transition is assumed to be at a constant current if the energy stored in the inductor is much higher than the energy in the capacitors:

$$\frac{1}{2}LI^2 \gg q_{oss\_Total}(V_{IN}) \cdot V_{IN} \tag{12}$$

where $q_{oss\_Total}(V_{IN})$ is the total charge of the output capacitance of a switch. A switch may consist of several transistors in parallel. To avoid confusion with the quality factors, which are traditionally denoted with an uppercase Q, all the charges for the ZVS analysis are denoted with a lowercase q. If the condition in Equation (11) holds, then the turn-off current $i_{off}(t)$ can be assumed to be constant I_{off} and the charge delivered by the inductive current is the following, where t_d is the dead-time:

$$q_i = \int_0^{t_d} i_{off}(t) \cdot dt \tag{13}$$

The complete ZVS condition can be derived in terms of available inductive charge (q_i) compared to the stored charge q_{oss_Total} using the following equation:

$$q_i \geq q_{oss\_Total}. \tag{14}$$

However, in LLC converters, the transition current is not constant, because multiple reactive elements ($L_r, L_m, C_r, \ldots$) shape the turn-off current, as illustrated in Figure 14. To account for this, a charge ratio k_q can be introduced:

$$k_q = \frac{q_i}{I_{off} \cdot t_d} = \frac{\int_0^{t_d} i_{off}(t) \cdot dt}{I_{off} \cdot t_d}, \tag{15}$$

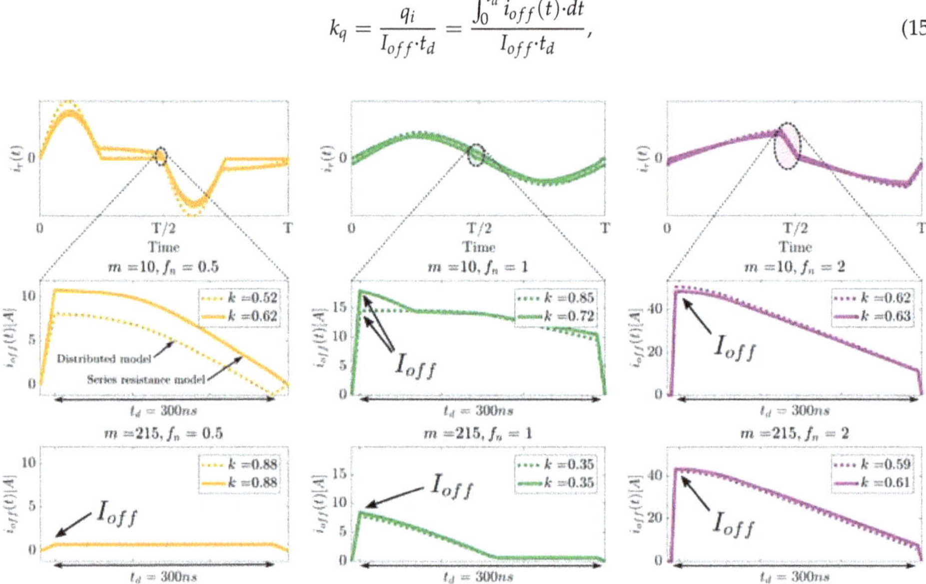

Figure 14. Time-based (TB) simulated currents of each mode and close-up of the dead-time transition for different LLC converter operation modes.

With this charge ratio, the relation between the initial turn-off current (I_{off}) and the real charge (q_i) can be defined for a given dead-time (t_d). When the turn-off transition occurs at constant current, the charge ratio has a value of $k_q = 1$. In LLC converters, when the turn-off current has a resonant shape, the value of the charge ratio is $k_q < 1$, as illustrated in Figure 14.

Complete ZVS transition is achieved when q_i is higher than q_{oss_Total} (see Equation (13)). For LLC converters, this condition can be estimated using TB models, as depicted in Figure 15. For $m = 215$, complete ZVS is achieved above resonance and depends on the output load. However, for $m = 10$, the complete ZVS range is greatly improved in the range of $0.6 < f_n < 1.5$ for the whole load range.

Using the information of Figure 15, it can be concluded that, for LLC converters, complete ZVS can only be achieved for the whole load range at low inductance ratios ($m = 10$). The best operation point is at resonant frequency, where $k_q \approx 1$ for all loads and the switching losses will be reduced. The low inductance ratio LLC converter can only achieve complete ZVS at heavy loads and operate above resonance when the efficiency is not optimal.

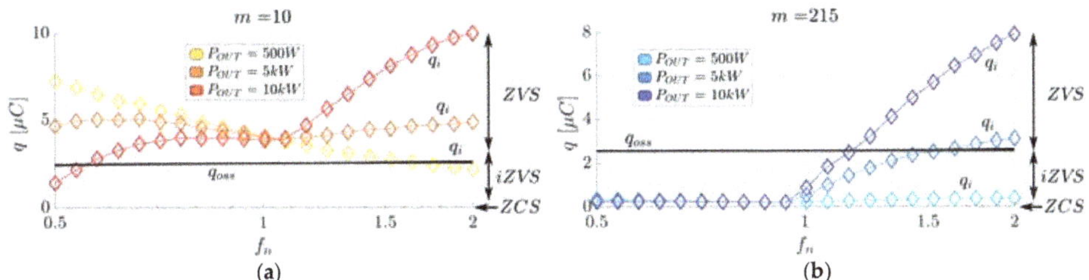

Figure 15. Time-based model simulations: inductive charge (q_i) and transistor output capacitor charge (q_{oss_Total}), where iZVS stands for incomplete ZVS. (**a**) for $m = 10$; (**b**) for $m = 215$.

4. Proposed Auxiliary Circuit

An auxiliary circuit is proposed to achieve complete ZVS for the full load range for a high inductance ratio (m), where the design of the LLC achieves the highest efficiency and power density. Auxiliary circuits are commonly used in full-bridge converters to improve their ZVS capabilities [16,28], particularly in LLC converters [29].

The auxiliary circuit consists of an inductor directly connected to both middle points of the primary full-bridge legs, as depicted in Figure 16. A small-series DC blocking capacitor (C_x) is added to avoid the core saturation of L_x, since the inductor current is not controlled.

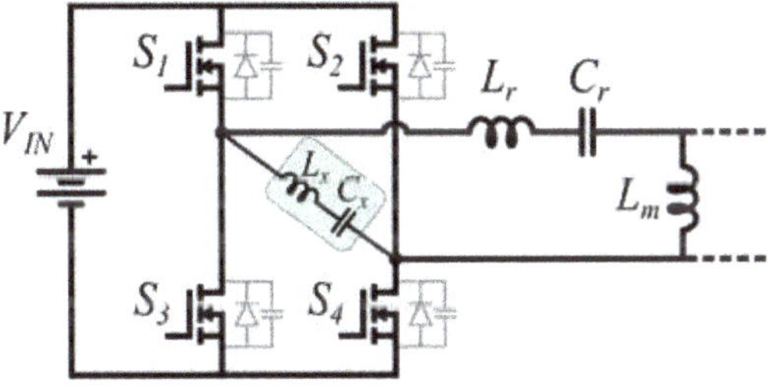

Figure 16. Proposed auxiliary circuit.

The purpose of the auxiliary inductor is to provide a constant current during the ZVS charge/discharge transition, avoiding the issues discussed in the previous section.

The equivalent circuit for FHA analysis is modified as illustrated in Figure 17.

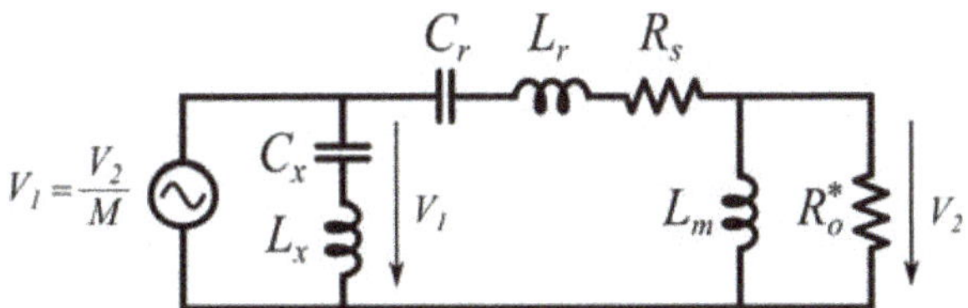

Figure 17. FHA equivalent circuit with auxiliary circuit.

Because the auxiliary circuit is in parallel with the input voltage source, the gain will not be affected by this circuit. Additionally, the resonant frequency of the auxiliary circuit is designed to be much lower than the series resonant tank frequency in order to not affect the series resonant tank frequency. To analyze this circuit, two additional normalized parameters need to be introduced:

- The normalized auxiliary frequency:

$$f_{xn} = \frac{f_x}{f_r} = \sqrt{\frac{L_r C_r}{L_x C_x}} \qquad (16)$$

- The auxiliary inductance ratio m_x:

$$m_x = \frac{L_x + L_r}{L_r} \qquad (17)$$

For simplicity, these parameters are defined similarly to the normalized parameters of the LLC converter.

4.1. Auxiliary Circuit Analysis and Design

Because the auxiliary circuit is designed to have an inductive behavior at the switching frequency ($f_x < f_s$), the capacitor voltage ripple can be neglected and the auxiliary inductor voltage is equal to the input bridge square voltage, as depicted in Figure 18.

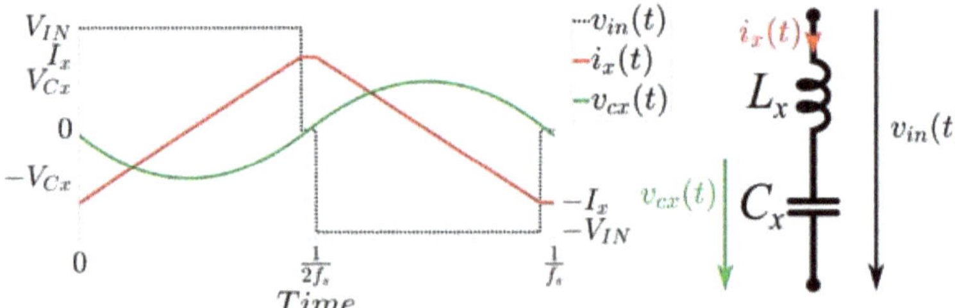

Figure 18. Auxiliary circuit current and voltage and auxiliary capacitor voltage ripple.

The triangular inductive peak current can be calculated as follows:

$$I_x = max(i_x(t)) = \frac{V_{IN}d}{2L_x f_s} = \frac{\pi d n V_{OUT}}{(m_x - 1)Z_r f_n M}, \qquad (18)$$

where d is the duty cycle, and Z_r is the equivalent impedance of the series resonant tank.

As explained in the previous section, Equation (13) must hold to achieve complete ZVS. However, with the auxiliary circuit, the total charge is equal to the charge delivered by the LLC tank current and the auxiliary circuit current.

$$q_i = q_{ix} + q_{ir} = k_q I_{off} t_d = k_q (I_x + I_r) t_d. \qquad (19)$$

Using Equations (13), (17) and (18), the maximum value for the inductance ratio of the auxiliary circuit m_x to ensure complete ZVS for a given t_d can be calculated as follows:

$$m_x \le 1 + \frac{\pi d n V_{OUT}}{\left(\frac{q_{oss}}{k_q \cdot t_d} - I_r\right) Z_r f_n M}. \qquad (20)$$

As this auxiliary circuit can provide almost all of the charge to achieve complete ZVS, the LLC can be designed with a high magnetizing inductance value to minimize the circulating currents and optimize efficiency, as explained in the first section.

The maximum auxiliary inductance ratio from Equation (19) is plotted at different loads and frequencies in Figure 19. The selected auxiliary inductance ratio for the final prototype is 12.5.

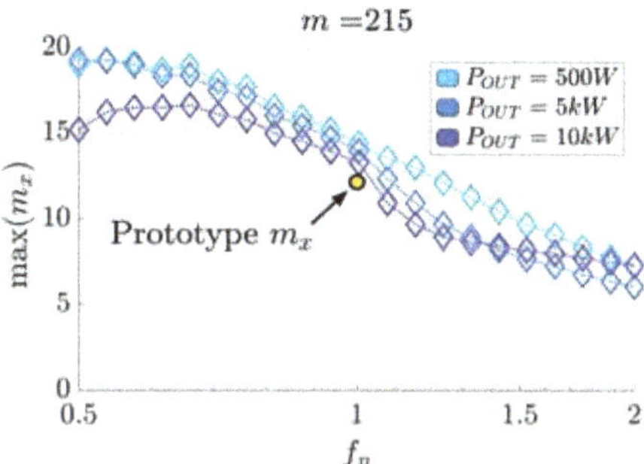

Figure 19. Auxiliary circuit maximum inductance ratio (m_x) design criteria.

To avoid DC bias in the unregulated inductor, a DC blocking capacitor is added in series. This creates a secondary resonant tank. As explained previously, the auxiliary circuit behavior is inductive. At resonant frequency, where $f_s = f_r$, then $f_{xn} \ll 1$.

The voltage in the capacitor is negligible compared to the input square waveform. However, the maximum voltage value should be calculated for design purposes. The auxiliary capacitor voltage shape is a parabola, as shown in Figure 18. The ripple in this capacitor can be calculated using the capacitor equation and integrating the triangular current through the auxiliary circuit:

$$V_{Cx} = max(v_{Cx}(t)) = v_{Cx}\left(\frac{d}{2f_s}\right) = \frac{I_x\left(1 - \frac{3}{2}d\right)}{2C_x f_s} = (\pi)^2 \cdot d\left(1 - \frac{3}{2}d\right) \cdot \frac{f_{xn}^2}{f_n^2} \cdot \frac{nV_{OUT}}{M} \tag{21}$$

For a value of $f_{xn} = 0.1$, the equation yields $V_{Cx} \approx 0.02 \cdot V_{IN}$.

4.2. Transformer Design and Comparison

Using the auxiliary circuit, complete ZVS at a full load range can be achieved without using the magnetizing inductance of the transformer. Therefore, the inductance ratio of the LLC converter can be as high as possible, reducing the circulating currents and improving the efficiency. To show the advantages of a high inductance ratio transformer, two transformers designs are compared.

The comparison between the $m = 10$ transformer design, Figure 20, and the $m = 215$ transformer with $m_x = 10$, Figure 21, auxiliary inductor is shown in Table 3. The transformers are divided into two components because a series/parallel configuration is used; this will be explained in the next section.

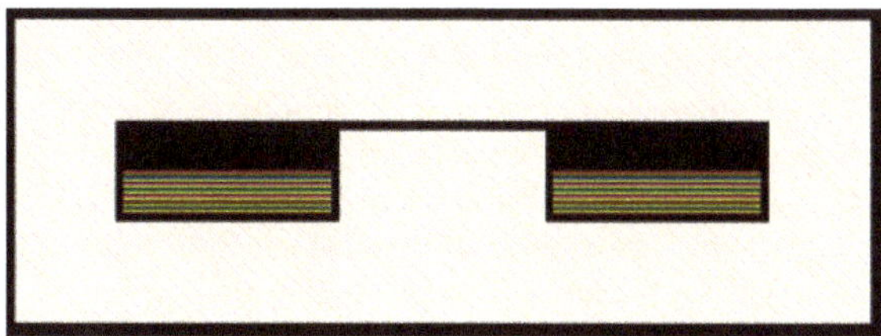

Figure 20. Design II: gapped transformer for $m = 10$, with a custom core and the same winding configuration and scale.

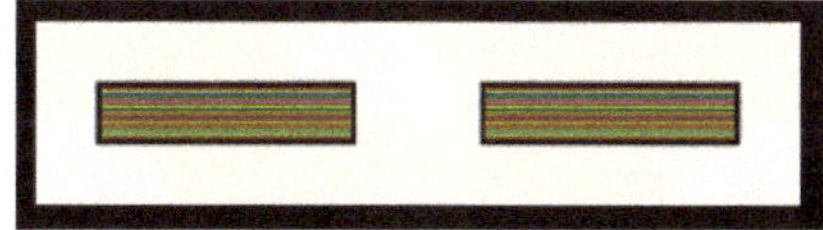

Figure 21. Design I: forward-type transformer for $m = 215$: planar E-core and 200 μm copper foil.

Table 3. Breakdown of losses, volume, and temperature of the two magnetic component designs.

	Transformers [1]				Auxiliary Inductor				Total Losses [1]	Total Volume [1]
	Power Loss		$\Delta\theta$	**Volume**	**Power Loss**		$\Delta\theta$	**Volume**		
	P_{Cu}	P_{Fe}			P_{Cu}	P_{Fe}				
Design I	9.8 W	13 W	61 °C	0.18 dm³	2.0 W	2.2 W	55 °C	0.13 dm³	49.8 W	0.49 dm³
Design II	25.5 W	18 W	101 °C	0.37 dm³	-	-	-	-	87 W	0.74 dm³

[1] There are two transformers in the design, as the final prototype uses a series parallel configuration. The total volume and losses include the two transformers.

Transformers with low magnetizing inductance typically use gaps and therefore lead to higher losses and reduced power density [25]. The window utilization of gapped transformers has to be low to avoid the gap proximity effect [30]. The proposed auxiliary circuit increases the magnetic component count but decreases the losses and volume. It can be concluded that the proposed auxiliary circuit is an overall improvement to the power density of the converter and does not add any significant complexity to the control scheme, as it is a passive solution.

5. Prototype

A prototype was built to validate the proposed solution for a 10 kW aircraft application with the parameters shown in Tables 4 and 5.

Table 4. Normalized parameters of the prototype.

Q (500 W)	Q (5 kW)	Q (10 kW)	Q_S	m	m_x	f_n	f_{xn}
0.02	0.18	0.36	7.5	215	12.5	0.99	0.023

Table 5. Parameters of the prototype.

L_r	C_r	L_m	L_x	C_x	R_s	f_r	f_s
7.11 µH	349 nF	2×750 µH	74 µH	60 µF	602 mΩ	102 kHz	101 kHz

A modified LLC converter topology was used with two series parallel transformers and two output bridges [31], as depicted in Figure 22. This modified structure does not affect the analysis described in previous sections. However, the series inductor L_r of the LLC topology consists on the leakage inductances of both transformers in series (L_{rA} and L_{rB}), while the parallel inductor L_m consists on the two transformers magnetizing inductances in series (L_{mA} and L_{mB}). Because the magnetizing inductances of the two transformers are in series in the primary mode, the inductance ratio (m) is further increased.

This modified structure is adequate for high-power applications since the two independent output bridges can handle half the load and double the switching devices effectively in parallel. Synchronous output bridges are used to further improve the efficiency. Current sharing in the secondary mode is achieved because of the series connection of the two transformers. Therefore, current measurement or a secondary resonant capacitor are not required. Because the difference in the impedance of each secondary bridge is small, the mismatch in the current sharing will also be small. However, secondary bridge synchronization is important to ensure a balance of DC bias in both transformers; synchronization can occur naturally with higher dead-times and under diode turn-off [5].

The prototype is depicted in Figure 23 and consists of the input board, the transformers, and the output board, as well as the control board, which provides all transistor signals as well as providing hardware protection.

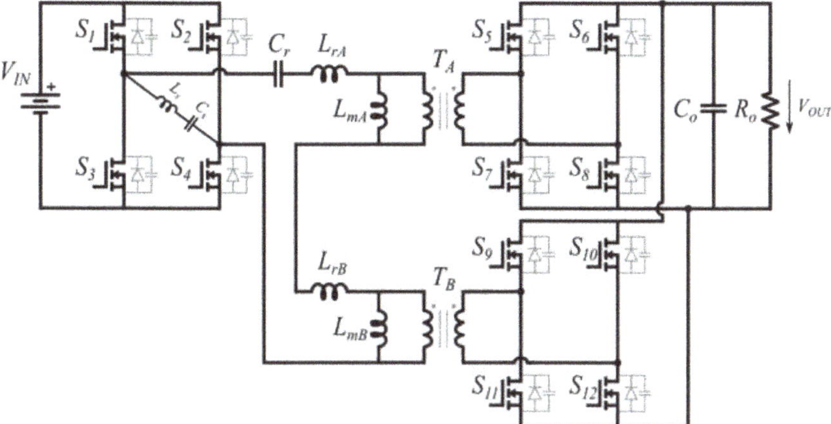

Figure 22. LLC full-bridge with a series parallel transformer configuration and two output-synchronous full-bridges.

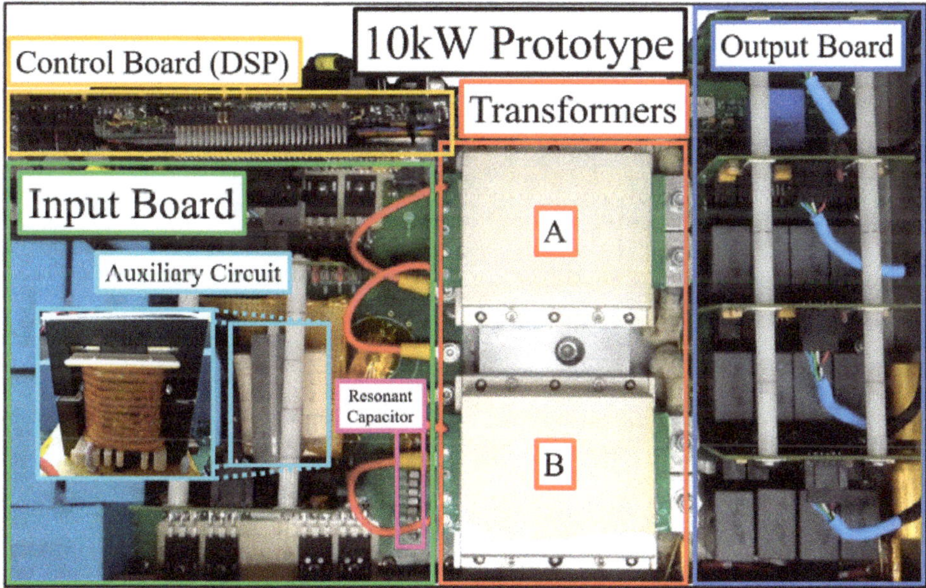

Figure 23. 10 kW prototype.

The components used in the prototype are shown in Table 6. Silicon MOSFET technology is used because it is a robust technology. Because the converter operates with soft-switching, the use of an Si devices does not detract from the efficiency. However, improvements in other types of devices, such as high-voltage GaN technology, could be a candidate for future improvements on the prototype. These devices will further reduce the volume of the auxiliary circuit as they have less output capacitance charge and ZVS is achieved more easily.

Table 6. Prototype components.

Component	Reference	Type	Quantity
Input capacitor	B32778G8606K	Film	5
Primary transistors	IPW65R037C6	Coolmos	8
Resonant capacitor	C4532C0G2E473J320KA	Multilayer Ceramic	6
	C3225C0G2E103J160KA		3
Transformer core	Planar E-core	Ferrite	2
Transformer winding	Copper foil	200 μm	-
Secondary transistors	IPP10004S2L-03	Optimos	32
Output capacitors	B32774D4226J000	Film	12
Auxiliary inductor core	E65/32/27	Ferrite	1

6. Experimental Results

Experimental waveforms at the nominal and overload power (5 kW and 10 kW) are depicted in Figure 24. The secondary currents are measured with the inverted polarity of one measurement as they fully overlap and would not otherwise be distinguishable. It can be seen that the current sharing is validated, with a DC mismatch of less than 5% at 10 kW.

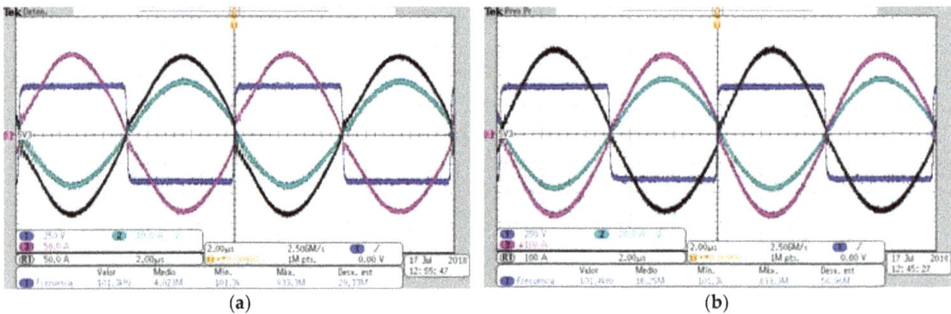

(a) (b)

Figure 24. Experimental results: secondary transformer current (magenta and black), input resonant current (cyan), input voltage bridge (blue); (**a**) at 5 kW of output power; (**b**) at 10 kW of output power.

The measurements are depicted in Figure 25a. The experimental results of the ZVS transition are shown in Figure 25b at 500 W (light load), in Figure 26a at 5 kW (nominal load) and in Figure 26b at 10 kW (overload condition). The ZVS transition is achieved completely for all loads as the drain-source voltage of S3 is zero when it is turned on. The charge ratio can be estimated integrating the current between the cursors $k_q = 0.91/0.66/0.55$ for 500 W, 5 kW and 10 kW, respectively.

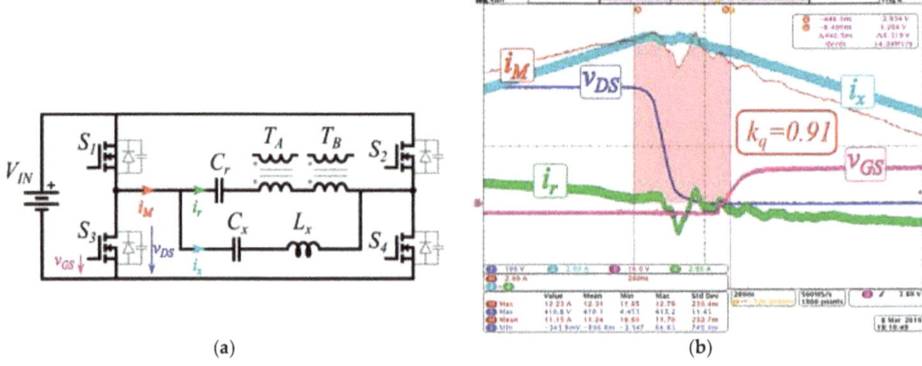

(a) (b)

Figure 25. (**a**) Measurement setup including auxiliary circuit; (**b**) ZVS experimental results at 500 W.

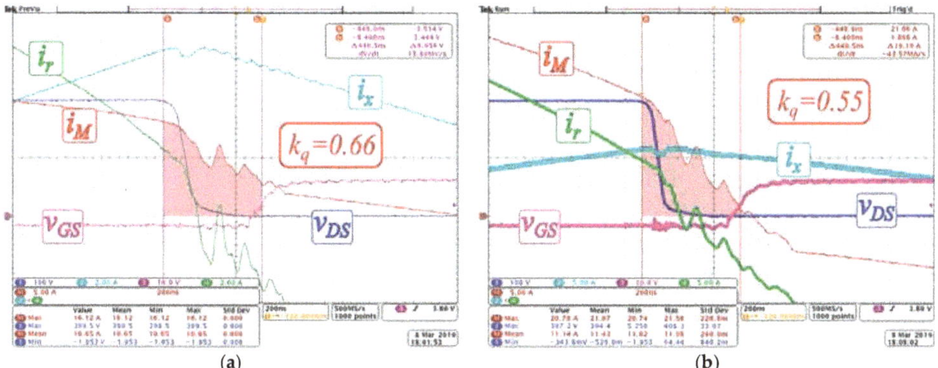

(a)   (b)

Figure 26. ZVS experimental results: (**a**) at 5 kW (**b**) at 10 kW.

To validate the improvement in efficiency provided by the auxiliary circuit, prototype tests are carried out without the auxiliary circuit. In Figure 27, the resonant current, drain-source and gate-source voltages are shown. ZVS is not achieved and voltages spikes appear in both voltages. The resonant current is distorted by a parasitic resonance between the output capacitor of the MOSFET and the series resonant inductor.

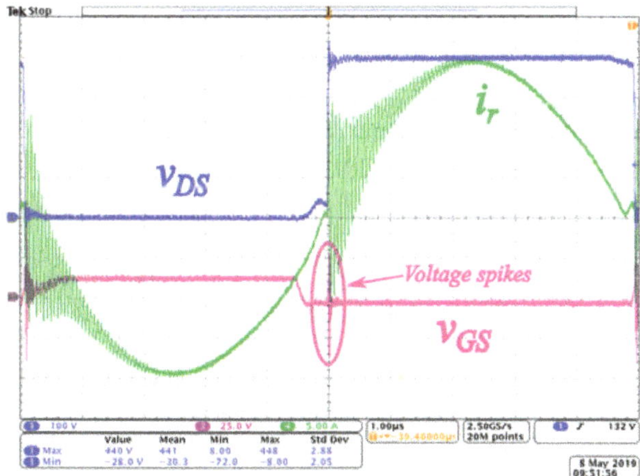

Figure 27. Experimental results without auxiliary circuit at 5 kW.

The performance comparison between these two prototype set-ups is depicted in Figure 28a. The efficiency is improved by the inclusion of the auxiliary circuit. Without the auxiliary circuit, ZVS is not achieved and the switching losses are increased.

A close-up of the measured efficiency at different loads is depicted in Figure 28a. The measured efficiency at nominal output power (5 kW) is 96.26% and at the overload power (10 kW) is 94.86%. The estimated efficiency is calculated using the power loss model explained in Section 2. The results show a good match with the estimated values.

The measured load-dependent voltage drop of the gain (M) is shown in Figure 28b and follows the modified analytical equation (Equation (6)).

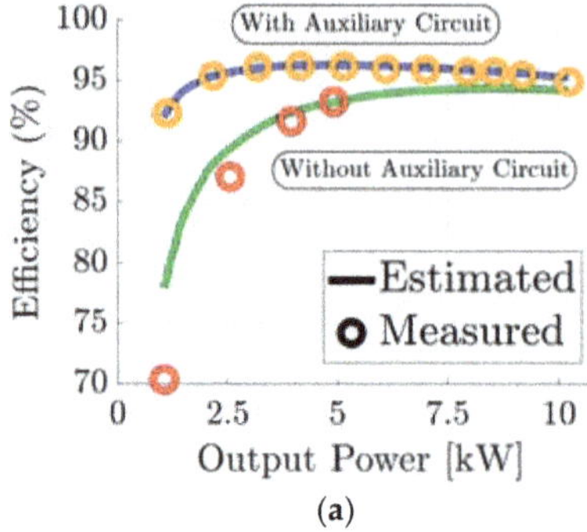

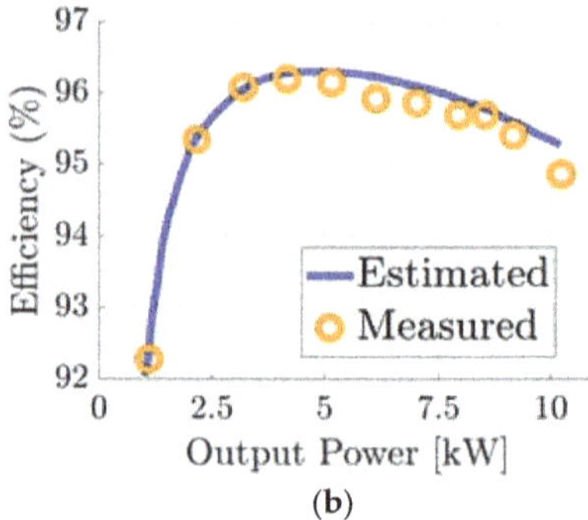

Figure 28. Experimental results: (**a**) efficiency at different output powers with (blue and yellow) and without (green and red) the auxiliary circuit, (**b**) close-up of the efficiency with the auxiliary circuit.

7. Conclusions

This article makes two main contributions: the first is related to the accurate modeling of LLC converters. A series resistance FHA model and a distributed FHA model are proposed and compared with time-based models and are an improvement over the traditional FHA approach. The first model is accurate for high inductance ratio LLC converters, while the second model is more accurate for low inductance ratios. The distributed FHA model has not been experimentally validated, and this can be done as future work. A time-based ZVS analysis is performed using a method based on charge and using time-based models. This analysis considers the variable current along the transition in LLC converters. With this method, the ZVS range is analyzed for different LLC designs. As a conclusion, the most suitable design to achieve ZVS in the whole load range is the low inductance ratio LLC converter, although it is less efficient than the high inductance ratio converter.

As for the second contribution, an auxiliary circuit is proposed to achieve complete ZVS at the whole load range in a high inductance ratio LLC converter. The circuit is analyzed using the charge method proposed. This circuit is designed for an unregulated LLC converter operating at resonant frequency. This solution has higher power density compared with the low inductance ratio LLC converter as it avoids gapped transformer designs.

A 10 kW prototype is designed to validate the auxiliary circuit. The measured efficiency is 96.2% at the 5 kW nominal power and 94.86% at the 10 kW overload power. Experimental results show that complete ZVS is achieved with the auxiliary circuit, at a light load (500 W), nominal load (5 kW) and peak load (10 kW). With the addition of the auxiliary circuit, the efficiency is improved from 93% to 96.2% at 5 kW.

Author Contributions: Conceptualization, Y.E.B. and D.S.; data curation, Y.E.B.; formal analysis, Y.E.B. and D.S.; funding acquisition, J.A.C. and J.C.; investigation, Y.E.B. and P.A.; methodology, Y.E.B., J.A.O. and P.A.; project administration, M.V., J.A.O., P.A. and J.C.; resources, G.M. and J.C.; software, Y.E.B.; supervision, M.V., J.A.O. and P.A.; validation, Y.E.B., U.B. and G.M.; visualization, Y.E.B.; writing—original draft, Y.E.B.; writing—review & editing, Y.E.B., D.S. and P.A.

Funding: This research has been partially supported by the research project DPI2017-88505-C2-1-R (Soluciones optimas para el procesado de la energia de las nuevas plantas solares de 1500V con conexion a la red electrica) funded by the Spanish government (Ministerio de Ciencia, Innovacion y Universidades).

Acknowledgments: The authors thank Isabel Senent and Cheng Li for the help in the review process, and the AIR-Project (E13-0505C-366) for contributing to build the prototype.

Conflicts of Interest: The authors declare no conflict of interest. The funders had no role in the design of the study; in the collection, analyses, or interpretation of data; in the writing of the manuscript, or in the decision to publish the results.

References

1. Emadi, K.; Ehsani, M. Aircraft power systems: Technology, state of the art, and future trends. *IEEE Aerosp. Electron. Syst. Mag.* **2000**, *15*, 28–32. [CrossRef]
2. Rosero, J.A.; Ortega, J.A.; Aldabas, E.; Romeral, L. Moving towards a more electric aircraft. *IEEE Aerosp. Electron. Syst. Mag.* **2007**, *22*, 3–9. [CrossRef]
3. Jones, R.I. The more electric aircraft: The past and the future? In Proceedings of the IEE Colloquium Electrical Machines and Systems for the More Electric Aircraft, London, UK, 9 November 1999; p. 1.
4. Zhao, X.; Guerrero, J.M.; Wu, X. Review of aircraft electric power systems and architectures. In Proceedings of the 2014 IEEE International Energy Conference (ENERGYCON), Cavtat, Croatia, 13–16 May 2014; pp. 949–953.
5. Bouvier, Y.E.; Borović, U.; Vasić, M.; Oliver, J.A.; Alou, P.; Cobos, J.A.; Árevalo, F.; García-Tembleque, J.C.; Carmena, J. DC/DC fixed frequency resonant LLC full-bridge converter with series-parallel transformers for 10 kW high efficiency aircraft application. In Proceedings of the 2017 IEEE Energy Conversion Congress and Exposition (ECCE), Cincinnati, OH, USA, 1–5 October 2017; pp. 3788–3795.
6. Borovic, U.; Zhao, S.; Silva, M.; Bouvier, Y.E.; Vasić, M.; Oliver, J.A.; Alou, P.; Cobos, J.A.; Árevalo, F.; García-Tembleque, J.C.; et al. Comparison of three-phase active rectifier solutions for avionic applications: Impact of the avionic standard DO-160 F and failure modes. In Proceedings of the 2016 IEEE Energy Conversion Congress and Exposition (ECCE), Milwaukee, WI, USA, 18–22 September 2016; pp. 1–8.
7. Bell, B.; Hari, A. Topology key to Power Density in Isolated DC-DC Converters. Available online: https://pdfs.semanticscholar.org/fbac/b2391b4df41435779e2c69690ef052db2481.pdf (accessed on 14 May 2019).
8. Yeon, C.-O.; Park, M.-H.; Ko, S.-H.; Lim, C.-Y.; Jang, Y.-J.; Moon, G.-W.; Kang, F.-S. A new LLC resonant converter with resonant frequency change for high conversion efficiency and high power density. In Proceedings of the 2017 IEEE 3rd International Future Energy Electronics Conference and ECCE Asia (IFEEC 2017-ECCE Asia), Kaohsiung, Taiwan, 3–7 June 2017; pp. 265–269.
9. Fei, C.; Lee, F.C.; Li, Q. High-efficiency high-power-density LLC converter with an integrated planar matrix transformer for high-output current applications. *IEEE Trans. Ind. Electron.* **2017**, *64*, 9072–9082. [CrossRef]
10. Pavlovic, Z.; Oliver, J.A.; Alou, P.; García, Ó.; Cobos, J.A. Bidirectional multiple port dc/dc transformer based on a series resonant converter. In Proceedings of the 2013 Twenty-Eighth Annual IEEE Applied Power Electronics Conference and Exposition (APEC), Long Beach, CA, USA, 17–21 March 2013.
11. Tomas-Manez, K.; Zhang, Z.; Ouyang, Z. Unregulated series resonant converter for interlinking DC nanogrids. In Proceedings of the 2017 IEEE 12th International Conference on Power Electronics and Drive Systems (PEDS), Honolulu, HI, USA, 12–25 December 2017; pp. 647–654.
12. Gopiyani, A.; Patel, V.; Shah, M.T. A novel half-bridge LLC resonant converter for high power DC power supply. In Proceedings of the 2010 Conference Proceedings IPEC, Singapore, 27–29 October 2010; pp. 34–39.
13. Hu, S.; Deng, J.; Mi, C.; Zhang, M. LLC resonant converters for PHEV battery chargers. In Proceedings of the 2013 Twenty-Eighth Annual IEEE Applied Power Electronics Conference and Exposition (APEC), Long Beach, CA, USA, 17–21 March 2013; pp. 3051–3054.
14. Fu, D.; Lu, B.; Lee, F.C. 1MHz high efficiency LLC resonant converters with synchronous rectifier. In Proceedings of the 2007 IEEE Power Electronics Specialists Conference, Orlando, FL, USA, 17–21 June 2007; pp. 2404–2410.
15. Zhang, X.; Chen, W.; Ruan, X.; Yao, K. A novel ZVS PWM phase-shifted full-bridge converter with controlled auxiliary circuit. In Proceedings of the 2009 Twenty-Fourth Annual IEEE Applied Power Electronics Conference and Exposition, Washington, DC, USA, 15–19 February 2009; pp. 1067–1072.
16. Safaee, A.; Jain, P.; Bakhshai, A. A robust low-RMS-current passive auxiliary circuit for ZVS operation of both legs in full bridge converters. In Proceedings of the 2015 IEEE Applied Power Electronics Conference and Exposition (APEC), Charlotte, NC, USA, 15–29 March 2015; pp. 21–27.
17. Chen, W.; Ruan, X.; Ge, J. A novel full-bridge converter achieving ZVS over wide load range with a passive auxiliary circuit. In Proceedings of the 2010 IEEE Energy Conversion Congress and Exposition, Atlanta, GA, USA, 12–16 September 2010; pp. 1110–1115.
18. Safaee, A.; Jain, P.K.; Bakhshai, A. An adaptive ZVS full-bridge DC–DC converter with reduced conduction losses and frequency variation range. *IEEE Trans. Power Electron.* **2015**, *30*, 4107–4118. [CrossRef]
19. Steigerwald, R.L. A comparison of half-bridge resonant converter topologies. *IEEE Trans. Power Electron.* **1988**, *3*, 174–182. [CrossRef]

20. Huang, H. FHA-based voltage gain function with harmonic compensation for LLC resonant converter. In Proceedings of the 2010 Twenty-Fifth Annual IEEE Applied Power Electronics Conference and Exposition (APEC), Palm Springs, CA, USA, 21–25 February 2010; pp. 1770–1777.
21. Liu, J.; Zhang, J.; Zheng, T.Q.; Yang, J. A modified gain model and the corresponding design method for an LLC resonant converter. *IEEE Trans. Power Electron.* **2017**, *32*, 6716–6727. [CrossRef]
22. Graovac, D.; Purschel, M.; Andreas, K. *MOSFET Power Losses Calculation Using the Data-Sheet Parameters*; Infineon Technologies AG: Neubiberg, Germany, 2006.
23. Li, J.; Abdallah, T.; Sullivan, C.R. Improved calculation of core loss with nonsinusoidal waveforms. In Proceedings of the Conference Record of the 2001 IEEE Industry Applications Conference. 36th IAS Annual Meeting (Cat. No.01CH37248), Chicago, IL, USA, 30 September–4 October 2001; Volume 4, pp. 2203–2210.
24. de Simone, S.; Adragna, C.; Spini, C. Design guideline for magnetic integration in LLC resonant converters. In Proceedings of the 2008 International Symposium on Power Electronics, Electrical Drives, Automation and Motion, Ischia, Italy, 11–13 June 2008; pp. 950–957.
25. Zhang, J.; Hurley, W.G.; Wolfle, W.H. Gapped transformer design methodology and implementation for LLC resonant converters. *IEEE Trans. Ind. Appl.* **2016**, *52*, 342–350. [CrossRef]
26. Yang, B.; Chen, R.; Lee, F.C. Integrated magnetic for LLC resonant converter. In Proceedings of the APEC. Seventeenth Annual IEEE Applied Power Electronics Conference and Exposition (Cat. No.02CH37335), Dallas, TX, USA, 10–14 March 2002; pp. 346–351.
27. Kasper, M.; Burkat, M.R.; Deboy, G.; Kolar, J.W. ZVS of power MOSFETs revisited. *IEEE Trans. Power Electron.* **2016**, *31*, 8063–8067. [CrossRef]
28. Borage, M.; Tiwari, S.; Kotaiah, S. A passive auxiliary circuit achieves zero-voltage-switching in full-bridge converter over entire conversion range. *IEEE Power Electron. Lett.* **2005**, *3*, 141–143. [CrossRef]
29. Kim, D.-K.; Moon, S.; Yeon, C.-O.; Moon, G.-W. High efficiency LLC resonant converter with high voltage gain using auxiliary LC resonant circuit. *IEEE Trans. Power Electron.* **2016**, *31*, 6901–6909. [CrossRef]
30. Prieto, R.; Cobos, J.A.; Garcia, O.; Alou, P.; Uceda, J. High frequency resistance in flyback type transformers. In Proceedings of the APEC 2000. Fifteenth Annual IEEE Applied Power Electronics Conference and Exposition (Cat. No.00CH37058), New Orleans, LA, USA, 6–10 February 2000; Volume 2, pp. 714–719.
31. Lin, B.R.; Chen, P.L.; Huang, C.L. Analysis of LLC converter with series-parallel connection. In Proceedings of the 2010 5th IEEE Conference on Industrial Electronics and Applications, Taichung, Taiwan, 15–17 June 2010; pp. 346–351.

Article

Design Methodology of Tightly Regulated Dual-Output LLC Resonant Converter Using PFM-APWM Hybrid Control Method [†]

HwaPyeong Park, Mina Kim, HakSun Kim and JeeHoon Jung *

Ulsan National Institute of Science and Technology (UNIST), Ulsan KS017, Korea; darkrla6@unist.ac.kr (H.P.); kmaop44@unist.ac.kr (M.K.); hszic@unist.ac.kr (H.K.)
* Correspondence: jhjung@unist.ac.kr or jung.jeehoon@gmail.com; Tel.: +82-052-217-2140
† This paper is an extended version of our paper published in Park, H.; Kim, M.; Jung, J. Tightly regulated dual-output half-bridge converter using PFM-APWM hybrid control method. In Proceedings of the IEEE Applied Power Electronics Conference and Exposition (APEC), Tampa, FL, USA, 26–30 March 2017; pp. 2022–2026.

Received: 15 May 2019; Accepted: 3 June 2019; Published: 4 June 2019

Abstract: A dual-output LLC resonant converter using pulse frequency modulation (PFM) and asymmetrical pulse width modulation (APWM) can achieve tight output voltage regulation, high power density, and high cost-effectiveness. However, an improper resonant tank design cannot achieve tight cross regulation of the dual-output channels at the worst-case load conditions. In addition, proper magnetizing inductance is required to achieve zero voltage switching (ZVS) of the power MOSFETs in the LLC resonant converter. In this paper, voltage gain of modulation methods and steady state operations are analyzed to implement the hybrid control method. In addition, the operation of the hybrid control algorithm is analyzed to achieve tight cross regulation performance. From this analysis, the design methodology of the resonant tank and the magnetizing inductance are proposed to compensate the output error of both outputs and to achieve ZVS over the entire load range. The cross regulation performance is verified with simulation and experimental results using a 190 W prototype converter.

Keywords: resonant converter; dual output converter; pulse frequency modulation (PFM); asymmetric pulse width modulation (APWM)

---

1. Introduction

Nowadays, many industry fields require well-regulated multiple output voltages to guarantee the stable operation of products, such as ultra-high-definition (UHD) TVs, computers, and other home appliances. To satisfy this requirement, point-of-use power supplies (PUPS) have been used for multiple output applications. However, this method has disadvantages of bulky size and low cost-effectiveness with many power converter modules [1]. Therefore, tightly regulated multiple output converters have been developed to improve the power density and the cost-effectiveness.

In previous research for multiple output converters, cross regulation methods have been popular, since they require output voltage sensors to obtain the output voltage regulation. However, wide load variations between the multiple output channels induce large output voltage error [2–7]. The secondary side post regulators (SSPR) have been proposed to tightly regulate the output voltage with small output voltage error. They can regulate each output voltage independently, however, additional switches, gate driving circuits, and voltage controllers are required [8–20].

In terms of topology, the LLC resonant converter is attractive for several applications, because it has soft switching capability and a small number of resonant components [21–24]. In previous research of the multiple output LLC resonant converter, the cross regulation technique has been used to regulate multiple output voltages [25,26]. In addition, the SSPR has been used to achieve tight output voltage regulation for the LLC resonant converter [27–30]. The LLC resonant converter using conventional control methods has the tradeoff between the cost effectiveness and regulation performance.

To obtain the high cost-effectiveness and tight output voltage regulation, the concept of a hybrid control method employing PFM and APWM was introduced for the dual-output LLC resonant converter in [31]. It does not require any additional components to implement the hybrid control algorithm, which shows the same cost-effectiveness as the conventional cross regulation method. However, it can only be applied to the dual-output converter. This previous research shows the preliminary operational principle and the decoupling algorithm to regulate the output voltages using the hybrid algorithm [32]. However, the previous research only shows the preliminary concept of the hybrid control algorithm. Therefore, the available voltage gain design, resonant tank design, and magnetizing inductance design are necessary to implement the hybrid control algorithm for the entire load condition with high power conversion efficiency.

In this paper, the design methodology of the dual-output LLC resonant converter with a hybrid control algorithm are proposed to regulate the output voltage with zero output voltage error and to obtain the ZVS capability for the entire load condition. The available voltage gain range is analyzed to implement the hybrid control algorithm. The design methodology of proper magnetizing inductance is proposed to achieve ZVS on the primary MOSFETs for the entire load range. The proper resonant tank design is proposed to compensate output voltage error for the worst-case load condition. In Section 4, the regulation performance of the dual-output LLC resonant converter with the hybrid control algorithm is verified through simulation and experimental results using a 190 W prototype converter.

2. Analysis of Dual-Output LLC Resonant Converter

The dual-output LLC resonant converter has the half-bridge structure of a primary inverting stage, a single transformer, and two output channels with diode rectifiers, as shown in Figure 1.

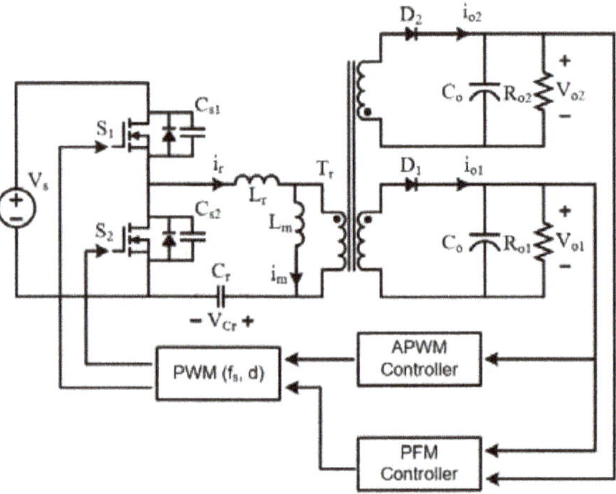

Figure 1. Schematic of the dual-output LLC resonant converter.

2.1. Operational Principle

Figure 2 shows the operation modes of the dual-output LLC resonant converter, which is divided into six modes during a single switching period. Mode 4 to Mode 6 are repeated from the previous half switching cycle. Figure 3 shows the operational waveforms of the dual-output converter. Mode 1 and 4 are a series resonant mode between the resonant inductance and the resonant capacitance. During Mode 1, the proposed converter transfers electric power in the primary side to V_{o1} in the secondary side. The primary and magnetizing current for Mode 1 can be derived as follows:

$$i_r(t) = I_{ini} \cos(\omega_r t) + C_r \omega_r (V_{in} - nV_{o1} - V_{cr,ini}) \sin(\omega_r t)$$
$$i_m(t) = I_{ini} + \frac{nV_{o1}}{L_m} t \tag{1}$$

where I_{ini} is the initial magnetizing current, C_r is the resonant capacitance, L_m is the magnetizing inductance, ω_r is the resonant angular frequency, V_{in} is the input voltage, n is the primary to secondary transformer turn ratio, V_{o1} is the one output voltage, and V_{cr} is the initial resonant capacitor voltage. During Mode 4, the converter transfers electric power to the V_{o2} side. The primary and magnetizing current for Mode 4 can be derived as follows:

$$i_r(t) = I_{ini}' \cos(\omega_r t) + C_r \omega_r (V_{cr,ini}' - nV_{o2}) \sin(\omega_r t)$$
$$i_m(t) = I_{ini}' - \frac{nV_{o2}}{L_m} t \tag{2}$$

where I_{ini}' is the initial magnetizing current, and $V_{cr,ini}'$ is the initial resonant capacitor voltage.

Mode 2 and 5 are parallel resonant modes among the resonant inductance, the magnetizing inductance, and the resonant capacitance. Those modes guarantee soft commutation on the secondary side diode rectifiers. Mode 3 and 6 are the dead time durations of the primary switch. During Mode 3 and 6, the output capacitance of the power switches is charged and discharged to obtain the ZVS. In steady state, the dual-output converter using the hybrid control algorithm can operate under a four operation mode (Case A), a five operation mode (Case B), and a six operation mode (Case C) according to output power conditions.

The light load condition makes the Case A operation, which induces no soft commutation on all output rectifiers during Mode 3 and 6. From middle to full load condition the converter operates according to Case B, which induces soft commutation on the diode of the V_{o1} channel during Mode 2 and 3. However, Mode 6 induces no soft commutation on the diode of the V_{o2} channel. When the same amount of power is transmitted to both output channels and the switching frequency is lower than the resonant frequency, the converter operates according to Case C, which induces soft commutation on all output rectifiers during Mode 2, 3, 5, and 6. Case C can show higher power conversion efficiency than that of Case A and B, because Case C has soft commutation capability on both the output rectifiers of V_{o1} and V_{o2} channels.

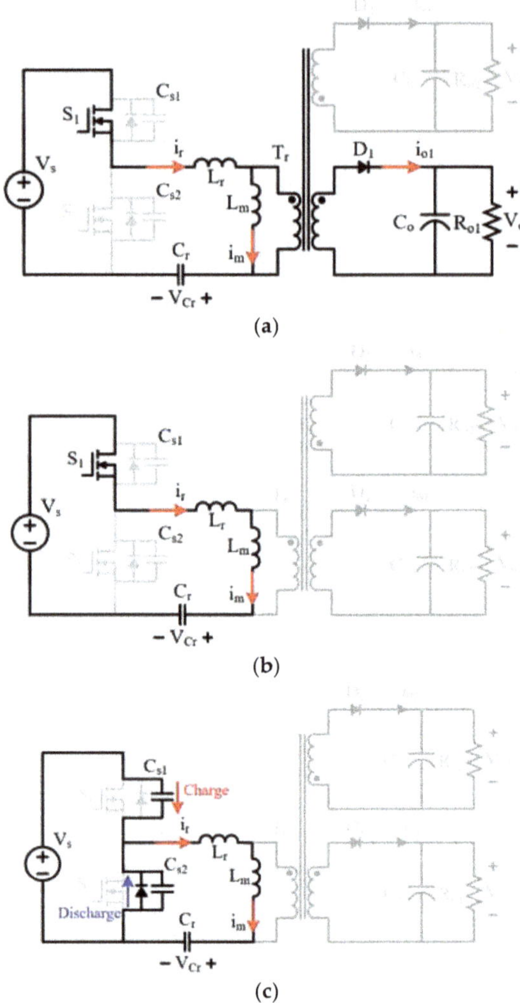

Figure 2. Operation mode of the dual-output LLC resonant converter: (**a**) Mode 1, (**b**) Mode 2, (**c**) Mode 3.

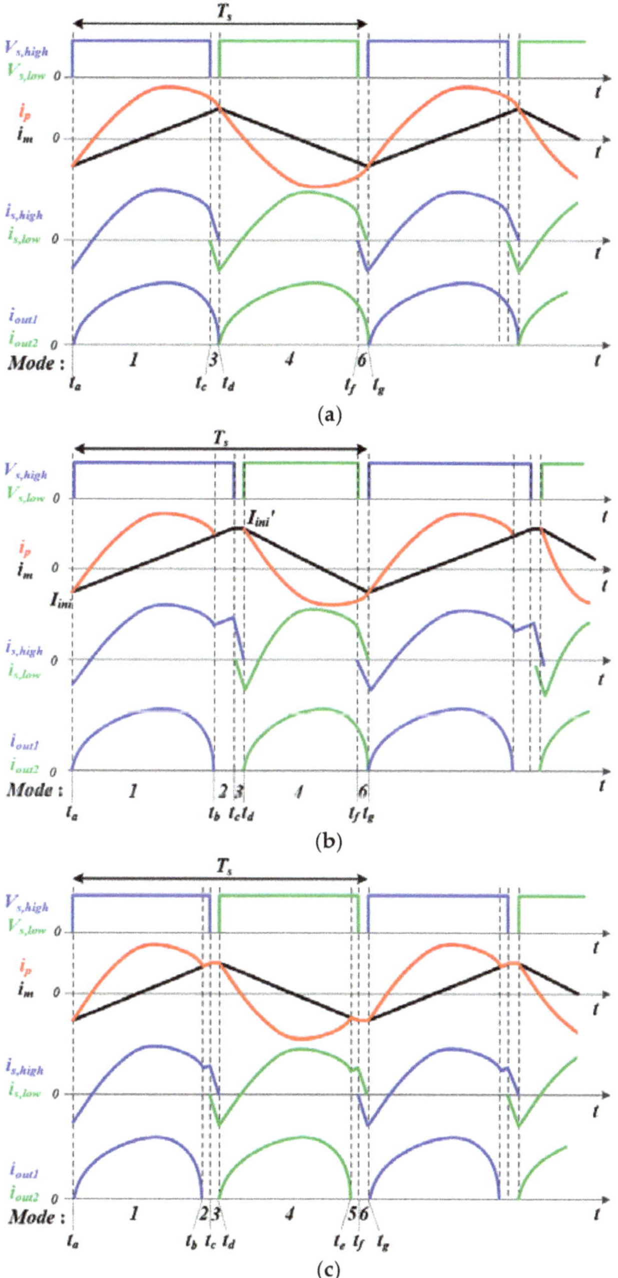

Figure 3. Operation mode of the dual-output LLC resonant converter: (**a**) Case A, (**b**) Case B, (**c**) Case C.

2.2. Gain Analysis According to Modulation Methods

The input-output voltage gain can be derived with the first harmonic approximation (FHA) as follows [29]:

$$H_r(f_n) = \left[\left(1 + k - \frac{k}{f_n^2}\right)^2 + Q^2\left(f_n - \frac{1}{f_n}\right)^2\right]^{-\frac{1}{2}} \tag{3}$$

where f_n is the normalized switching frequency, k is L_r/L_m inductance ratio, and Q is the quality factor as follows:

$$R_{o,e} = \frac{V_{o,FHA}}{I_{o,FHA}} = \frac{8n^2}{\pi^2}R_o, \; f_n = \frac{f_s}{f_r}, \; Q = \frac{\sqrt{\frac{L_r}{C_r}}}{R_{o,e}}, \; k = \frac{L_r}{L_m} \tag{4}$$

where R_o is the output resistance, f_s is the switching frequency, and f_r is the resonant frequency.

The conventional asymmetric half-bridge converter, only controlled by the APWM, uses small resonant inductance to obtain the linear voltage gain by the APWM [30]. However, it induces a flat voltage gain by the PFM, which cannot regulate the output voltage with PFM. Therefore, the resonant inductance is high enough to obtain the monotonic voltage gain by the PFM, as shown in Figure 4a. The large resonant inductance has limited monotonic voltage gain by the asymmetric duty variation.

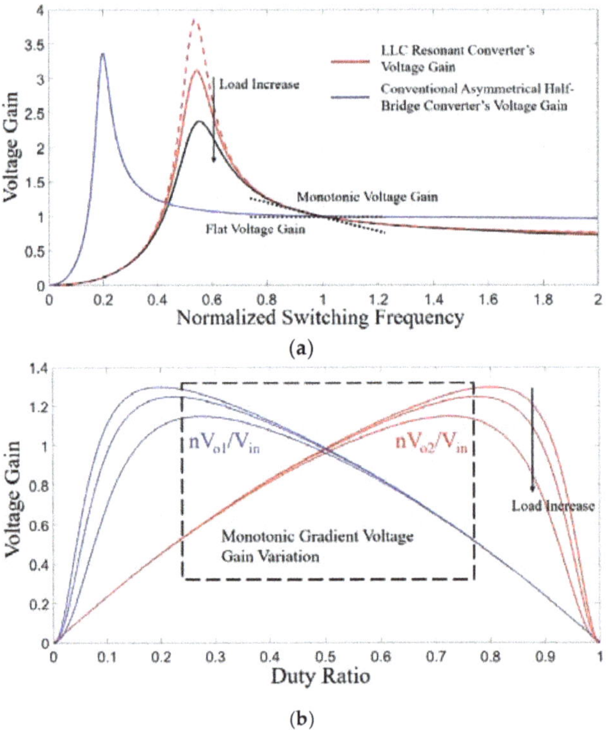

Figure 4. Voltage gain curves: (**a**) According to normalized switching frequency, (**b**) according to asymmetric duty ratio and load variation.

The conventional voltage gain of the APWM can be derived as follows [33]:

$$\frac{V_o}{V_{in}} \cong \frac{L_m D}{n(L_m + L_r)} \tag{5}$$

where D is the duty ratio. It has validity only for small resonant inductance conditions. Therefore, the voltage gain with enough resonant inductance is required to implement the hybrid control algorithm for the entire load conditions.

In steady state, the offset current on the magnetizing inductance can be derived as follows:

$$I_{offset} = \frac{I_{o1}}{n_1} - \frac{I_{o2}}{n_2} = \frac{1}{T_s} \int_0^{T_s} (i_r(t) - i_m(t))dt \tag{6}$$

Assuming that V_{cr} is constant in Mode 1, $V_{cr,ini}$ can be derived as follows:

$$\frac{nV_{o1}DT_s}{L_m} = \frac{V_{cr,ini}(1-D)T_s}{L_m}$$
$$V_{cr,ini} = \frac{1-D}{D}nV_{o1} \tag{7}$$

From (1), (2) and (6), (7), the proposed input to output voltage gains ($H_{o1}(D) = n_1 V_{o2}/V_{in}$, $H_{o2}(D) = n_2 V_{o2}/V_{in}$) according to the asymmetric duty ratio can be derived as follows:

$$\frac{nV_{o1}}{V_{in}} = \frac{D'\frac{1-\cos(A)}{Z_1\omega_1}}{\frac{D'}{\omega_r}\left(R'+\frac{DT_s}{2L_m}\right)\sin(A)+D'^2T_s\left(\frac{1}{n_1{}^2R_{o1}}-\frac{1}{n_2{}^2R_{o2}}\right)+\frac{1-\cos(A)}{Z_1\omega_r}}$$
$$\frac{nV_{o2}}{V_{in}} = \frac{D\frac{1-\cos(A')}{Z_1\omega_1}}{\frac{D}{\omega_r}\left(R'+\frac{(1-D)T_s}{2L_m}\right)\sin(A')+D^2T_sR'+\frac{1-\cos(A')}{Z_1\omega_r}} \tag{8}$$

where $D' = 1 - D$, $Z_1 = L_r/C_r$, $R' = 1/(n_1{}^2R_{o1}) - 1/(n_2{}^2R_{o2})$, $A = \omega_r DT_s$, and $A' = \omega_r D'T_s$.

Figure 4b shows the voltage gain of the APWM according to load variations. It shows a complementary voltage gain relationship between V_{o1} and V_{o2} in the monotonic gradient voltage gain region. From (8), the design of APWM operational range is necessary to obtain the monotonic gradient voltage gain at the designed APWM range.

2.3. Magnetizing Inductance Design for Soft Switching Capability

The ZVS capability of the primary MOSFETs is achieved by discharging and charging their output capacitance during the dead time. Therefore, the LLC resonant converter requires enough magnetizing current and dead time duration to guarantee ZVS condition which can be expressed as follows:

$$i_p(t_{dt}) \geq i_{req}(t_{dt}) \tag{9}$$

where i_{req} ($=2V_{in}C_s/t_{dt}$) is the required minimum primary current to obtain the ZVS condition on the primary MOSFETs, C_s is the equivalent output capacitance of the primary MOSFETs, and t_{dt} is the dead time duration.

From (6), the dual-output LLC resonant converter makes an unbalanced magnetizing current during the dead time based on each load condition of the dual-output. The unbalanced magnetizing current has to satisfy (9) to achieve ZVS operation. Therefore, the design of the magnetizing inductance and the dead time duration should take unbalanced magnetizing currents into consideration. Assuming the primary current is constant during the dead time, the proposed ZVS condition in the dual-output converter can be derived as follows:

$$t_{dt1} \geq \frac{(C_{s1}+C_{s2})V_{in}}{|i_p(t_c)|} = \frac{(C_{s1}+C_{s2})V_{in}}{\left|I_{off}+\frac{n_1V_{o1}}{2L_m}DT_s\right|}$$
$$t_{dt2} \geq \frac{(C_{s1}+C_{s2})V_{in}}{|i_p(t_c)|} = \frac{(C_{s1}+C_{s2})V_{in}}{\left|I_{off}-\frac{n_2V_{o2}}{2L_m}(1-D)T_s\right|} \tag{10}$$

where C_{s1} and C_{s2} are the output capacitance of $S1$ and $S2$, respectively, and $f_{s,max}$ is the maximum switching frequency, $t_{dt1} = t_d - t_c$ and $t_{dt2} = t_g - t_f$ are the first and second dead time of the primary MOSFETs, respectively. From (10), t_{dt} can be reformulated as follows:

$$t_{dt} \geq \frac{2C_s V_{in}}{\min\left\{\left|i_p(t_c)\right|, \left|i_p(t_f)\right|\right\}} \tag{11}$$

From (10) to (11), the proposed magnetizing inductance for the ZVS capability can be derived as follows:

$$L_m \leq \frac{t_{dt1} D_{min} T_s n_1 V_{o1}}{2\left(C_s V_{in} - I_{off,max} t_{dt1}\right)}$$
$$L_m \leq \frac{t_{dt2}(1-D_{max}) T_s n_2 V_{o2}}{2\left(C_s V_{in} - I_{off,max} t_{dt2}\right)} \tag{12}$$

From (11) and (12), the proposed magnetizing inductance and dead time duration can be designed for the dual-output LLC resonant converter to obtain ZVS capability of the primary MOSFETs over the entire load condition. These equations consider both the unbalanced magnetizing current and the switching frequency variation to achieve ZVS. Figure 5 shows the comparison of the required magnetizing inductance between the conventional LLC resonant converter and the dual-output converter to achieve ZVS. The dual-output converter requires lower magnetizing inductance for ZVS capability compared to the conventional LLC resonant converter.

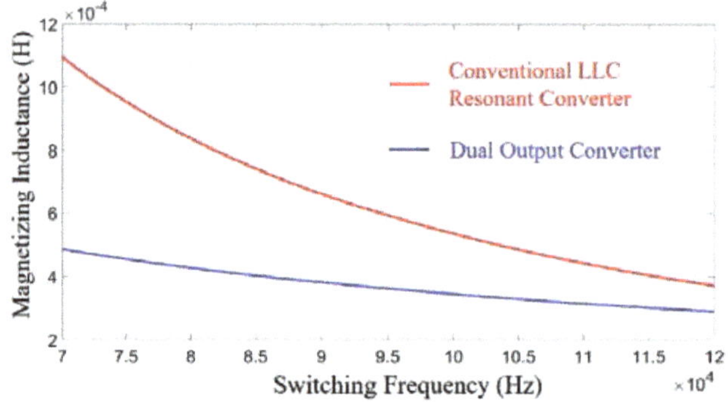

Figure 5. Magnetizing inductance according to the switching frequency for zero voltage switching (ZVS) capability.

3. Analysis of PFM and APWM Hybrid Control Algorithm and Resonant Tank Design

In this section, the operational principle of the hybrid control algorithm is analyzed to regulate output voltages. Through this analysis, the resonant tank is designed to achieve output voltage error compensation.

3.1. Analysis of the Hybrid Control Algorithm

The hybrid control algorithm has two control freedoms using two independent modulation methods. The PFM is adapted to regulate the output voltages using the conventional cross regulation method. In steady state, the conventional multiple output feedback can be derived as follows:

$$k_{w1} V_{o1} + k_{w2} V_{o2} = V_{ref} \tag{13}$$

where k_{w1} and k_{w2} are weight factors and V_{ref} is the reference output voltage. From (13), the output voltage errors using the weight factor can be derived as follows:

$$\begin{aligned}
k_{w1}\Delta V_{o1} + k_{w2}\Delta V_{o2} &= 0 \\
V_{o1} &= V_{o1,ref} + k_{w1}\Delta V_{o1} \\
V_{o2} &= V_{o2,ref} + k_{w2}\Delta V_{o2}
\end{aligned}$$

(14)

where $V_{o1,ref}$ and $V_{o2,ref}$ are the reference voltages of V_{o1} and V_{o2}, and ΔV_{o1} and ΔV_{o2} are output voltage errors of V_{o1} and V_{o2}, respectively. The feedback method for the multiple outputs affects the performance of the cross regulation using the weight factors. As a result, (14) cannot eliminate the output voltage errors which are divided into each output voltage according to the weight factor.

The APWM control method is adopted to regulate V_{o1}, which makes zero steady state voltage errors of the V_{o1} channel. The complementary voltage gain relationship between each output channel also reduces the output voltage error of V_{o2}, as shown in Figure 4b. The decrement of the V_{o1} and V_{o2} error using the APWM can be derived as follows:

$$\begin{aligned}
\Delta V_{o1}' &= H_{o1}(D) \cdot V_{o1} - V_{o1,ref} \cong 0 \\
\Delta V_{o2}' &= H_{o2}(D) \cdot V_{o2} - V_{o2,ref}
\end{aligned}$$

(15)

where $\Delta V_{o1}'$ and $\Delta V_{o2}'$ are the output voltage errors which are compensated with the APWM control method. The V_{o1} error is almost zero after the APWM control, however, the APWM cannot completely compensate the V_{o2} error. From (15), the output voltages after the APWM regulation can be derived as follows:

$$V_{o1}' = V_{o1,ref}, \quad V_{o2,ref} = V_{o2,ref} - \Delta V_{o2}'$$

(16)

where V_{o1}' and V_{o2}' are the output voltages of each output channel, which are compensated by the APWM. The flow chart of the hybrid control algorithm is shown in Figure 6 which shows the control sequence of the output voltage regulation to reduce the output voltage errors.

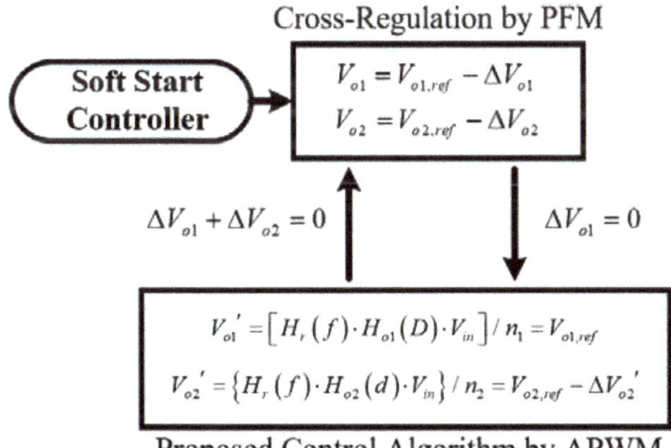

Figure 6. Block diagram of the PFM-APWM hybrid control algorithm.

After the APWM regulation, the cross regulation using the PFM with the feedback for the multiple outputs divides the output voltage error of V_{o2} with respect to the weight factors as follows:

$$\begin{aligned}
V_{o1,ref} - [H_{o1}(d)H_r(f_{s,c})V_{in}]/n_1 &\cong -k_{w2}\Delta V_{o2}' \\
V_{o2,ref} - [H_{o2}(d)H_r(f_{s,c})V_{in}]/n_2 &\cong -k_{w1}\Delta V_{o2}'
\end{aligned}$$

(17)

where $f_{s,c}$ is the switching frequency which divides the output voltage error according to the weight factor, and $H_r(f_{s,c})$ is the voltage gain of the PFM at the $f_{s,c}$. The voltage gain according to the switching frequency satisfies (13) to divide the output voltage error with respect to the weight factor. From (17), the output voltages regulated by the PFM can be derived as follows:

$$V_{o1}' = V_{o1,ref} - k_{w2}\Delta V_{o2}', \quad V_{o2,ref} = V_{o2,ref} + k_{w1}\Delta V_{o2}' \tag{18}$$

where V_{o1}'' and V_{o2}'' are the output voltages of each output channel, which are compensated by the PFM at the next control step. The control iterations of the PFM and the APWM can reduce the output voltage errors of V_{o1} and V_{o2}.

From (15) to (18), the voltage gain variation by the PFM and the APWM can decrease the output voltage error. However, the dual-output converter has limited voltage gain variation caused by the switching frequency and the duty ratio ranges. The proper operating ranges to obtain tightly regulated output voltages can be calculated as follows:

$$\begin{aligned} H_{o1}(d)H_r(f_s)V_1 = V_{o1,ref} \\ H_{o2}(d)H_r(f_s)V_2 = V_{o2,ref} \end{aligned} \tag{19}$$

where $V_1 = V_{in}/n_1 - I_{o1}R_{eff}$, $V_2 = V_{in}/n_2 - I_{o2}R_{eff}$, and R_{eff} is the effective series resistance. The voltage gain variation according to the PFM and the APWM has to satisfy (19) for the tight output voltage regulations. Using (19), the required voltage gain of the APWM can be calculated as follows:

$$H_{o1}(d) = \frac{V_{o1,ref}V_2}{V_{o2,ref}V_1}H_{o2}(d) \tag{20}$$

From (20), the APWM control method requires a voltage gain difference between $H_{o1}(d)$ and $H_{o2}(d)$ to reduce the output voltage error. The maximum voltage gain difference range of the APWM can be derived at the worst load condition as follows:

$$\frac{H_{o1}(d)}{H_{o2}(d)} = \max\left(\frac{V_{o1,ref}V_{2,min}}{V_{o2,ref}V_{1,max}}, \frac{V_{o1,ref}V_{2,max}}{V_{o2,ref}V_{1,min}}\right) \tag{21}$$

where $V_{1,max}$ and $V_{1,min}$ are minimum and maximum output voltages of V_{o1}, respectively, $V_{2,max}$ and $V_{2,min}$ are minimum and maximum output voltages of V_{o2}, respectively. These consider the voltage drop according to the forward bias and conduction loss. Figure 7a shows the maximum voltage gain difference between $H_{o1}(d)$ and $H_{o2}(d)$.

The required voltage gain of the PFM can be obtained for the cross regulation as follows:

$$H_r(f_s) = \frac{V_{o1,ref}}{H_{o1}(d)V_1} \tag{22}$$

From (22), the maximum and minimum voltage gains for the cross regulation can be derived as follows:

$$\begin{aligned} H_{r,max}(f_s) = \frac{V_{o1,ref}}{H_{o1,min}(d)V_{1,min}} \\ H_{r,min}(f_s) = \frac{V_{o1,ref}}{H_{o1,max}(d)V_{1,min}} \end{aligned} \tag{23}$$

where $H_{o1,min}(d)$ and $H_{o1,max}(d)$ are the minimum and maximum voltage gains of the APWM, respectively. Figure 7b describes the maximum and the minimum voltage gains to implement the cross regulation method. The resonant tank should be designed to compensate the maximum output voltage difference of the dual-output converter. The voltage gain according to the duty cycle and the switching frequency should satisfy (21) and (23) to compensate the maximum power difference. The power converter does not require any additional circuits to implement the hybrid control algorithm. Therefore, the power converter size is not affected by the control algorithm.

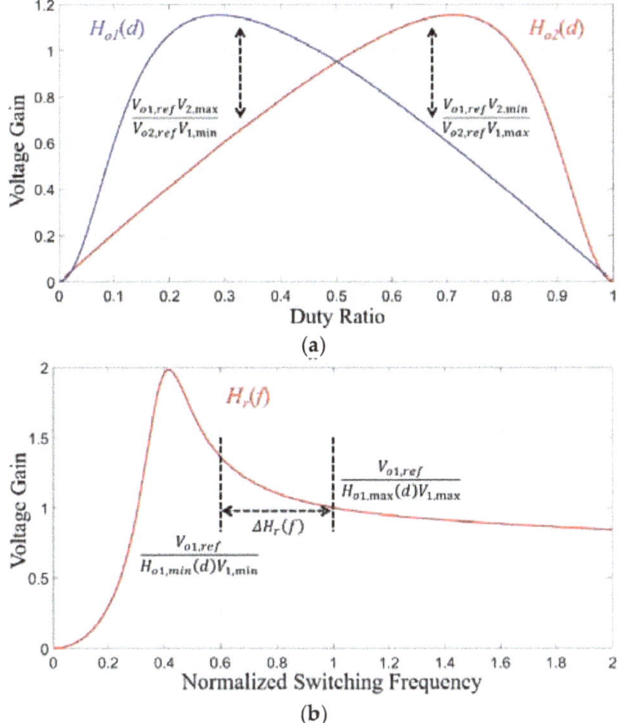

Figure 7. Required minimum and maximum voltage gains: (**a**) APWM case, (**b**) PFM case.

3.2. Resonant Tank Design for Minimizing Output Voltage Error

The voltage gain using APWM and PFM has to satisfy (21) and (23) for compensation of output voltage error over the entire load range. Figure 8 shows the voltage gain variation according to the resonant inductance and the modulation methods. Large resonant inductance makes monotonic voltage gain variation as switching frequency varies around the resonant frequency. However, the large resonant inductance reduces the compensation range according to the APWM by (8). On the other hand, small resonant inductance can compensate the large output voltage error using the APWM according to (5). However, it induces flat voltage gain variation according to the PFM, which makes large switching frequency variation to obtain the proper voltage gain for minimizing the output voltage error. Therefore, the maximum resonant inductance design is required to achieve small switching frequency variation and output voltage error compensation.

The maximum and minimum output voltages can be calculated with loss analysis. The primary side voltage drop can be calculated as follows:

$$V_{tr} = V_{in} - I_p(R_{ds} + R_{tr1} + R_c) \tag{24}$$

where V_{tr} is the transfer voltage from the primary to the secondary side, I_p is the primary current, R_{ds} is the on resistance of MOSFET, R_{tr1} is the primary resistance of the transformer, and R_c is the resistance of the resonant capacitor. In addition, the output voltage can be calculated as follows:

$$V_{o1} = V_{tr}/n - V_{D1} - I_{o1}R_{tr2}$$
$$V_{o2} = V_{tr}/n - V_{D2} - I_{o2}R_{tr2} \tag{25}$$

where V_{D1} and V_{D2} are the on-drop voltage of secondary diodes, I_{o1} and I_{o2} are the secondary rms currents of each output channel, and R_{tr2} is the series resistance of the transformer. Table 1 shows the specification of the proposed dual-output LLC resonant converter and its parasitic components.

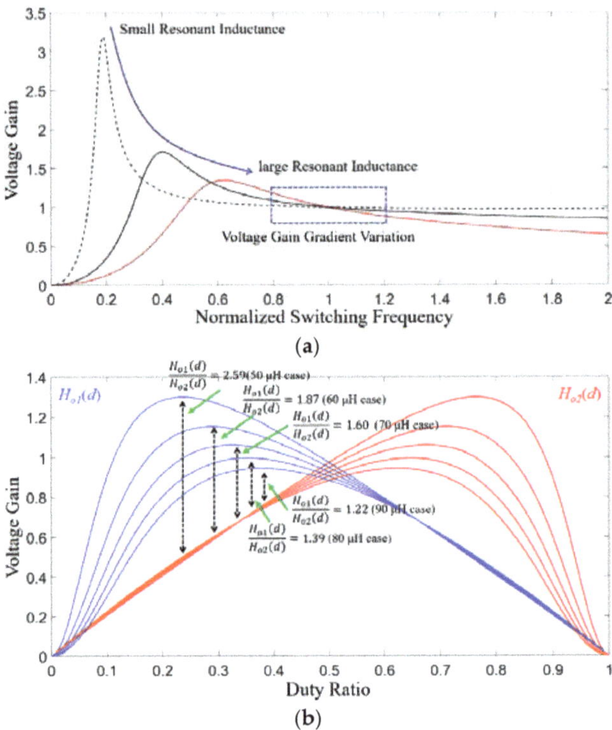

Figure 8. Voltage gain variation according to the resonant inductance: (**a**) PFM case, (**b**) APWM case.

Table 1. Power stage specifications and design parameters of the dual-output LLC resonant converter.

Parameter	Value	Parameter	Value	Parameter	Value
V_{in}	400 V	Load 1	20 V, 6 A	Load 2	10 V, 7 A
n_{mod}	12	L_m	380 μH	L_r	70 μH
C_r	30 nF	f_r	109 kHz	R_{ds}	330 mΩ
R_{tr1}	300 mΩ	R_{tr2}	130 mΩ	R_c	40 mΩ
$V_{D1,2}$	0.4 V				

From (8), (21) and (25), the resonant impedance (Z_1) can be designed to compensate the output voltage errors for the entire load range. Within the designed duty variation range, the resonant impedance can be derived as follows:

$$\sqrt{\frac{L_r}{C_r}} = \frac{x(D'B)\frac{B}{\omega_r} - (DB')\frac{B'}{\omega_r}}{(DB')C' - x(D'B)C} \tag{26}$$

where x is the maximum voltage gain ratio by (21), $B = 1 - \cos(A)$, $B' = 1 - \cos(A')$, $C = (D/\omega_r)[R' + (D'T_s)/2L_m]\sin(A') + D^2 T_s R'$, and $C' = (D'/\omega_r)[R' + (DT_s)/2L_m]\sin(A) + D'^2 T_s R'$. From (26), the resonant inductance and capacitance can be calculated to obtain the tight output voltage regulation.

The design methodology of the resonant tank can be described as follows. First, the specifications, such as input voltage range, resonant frequency, and output voltages, are required to design the power stage. Second, the magnetizing inductance is designed by considering the resonant frequency, the input voltage, and the parasitic capacitance of the MOSFETs, which was derived in (12). Third, the resonant inductance and capacitance ratio can be designed using (26). From the loss analysis, the required maximum voltage gain ratio can be calculated using (21), and (25). Through the proposed design methodology, the dual-output LLC resonant converter can achieve tight output voltage regulation and ZVS capability for the entire load conditions.

The example of the power stage design can be described as follows.

Step 1: The design specifications are shown in Table 1.

Step 2: The magnetizing inductance can be calculated with (12). The proper magnetizing inductance is 280 µH to achieve ZVS for the entire load range as shown in Figure 9.

Step 3: The required maximum voltage gain ratio is 1.31 and 0.64, which can be calculated with (21), and (25). The resonant inductance and capacitance can be calculated with (26), which are 70 µH and 30 nF. Therefore, the design example shows resonant impedance and magnetizing inductance to achieve ZVS for the entire load range, which can compensate all voltage errors of all the outputs using the hybrid control algorithm.

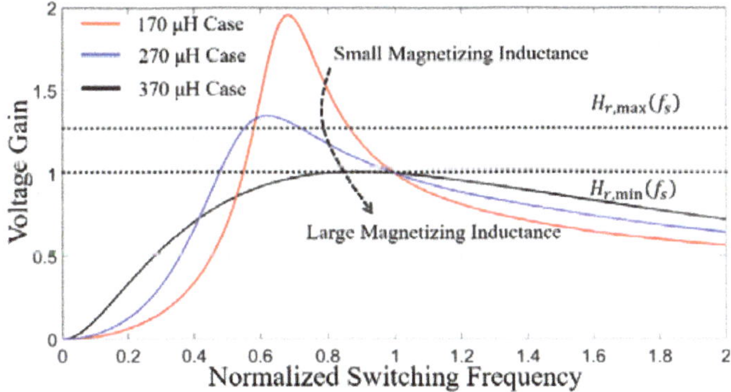

Figure 9. Compensation range of the output voltage error according to the resonant inductance.

4. Experimental Results

Figure 10 shows the simulation result of the PFM-APWM hybrid control method which has the minimum output voltage error, almost zero, according to the load variation. The simulation results verify the performance enhancement of tightly regulated output voltage of the hybrid control algorithm compared to the conventional cross regulation method. Figure 11 shows the prototype converter and diagram of the experimental setup to verify the performance of the proposed hybrid control algorithm. Figure 11a shows a photograph of the prototype converter which has two outputs and a single primary side. Two separated electronic loads are connected to the converter to verify the output voltage regulation performance as shown in Figure 11b.

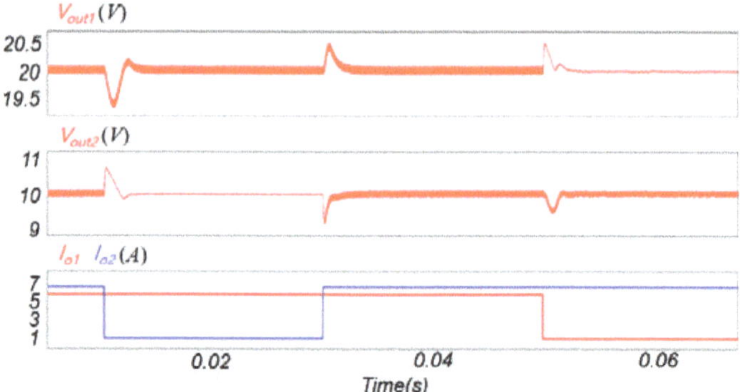

Figure 10. Simulation result using dual-output LLC converter with hybrid control algorithm.

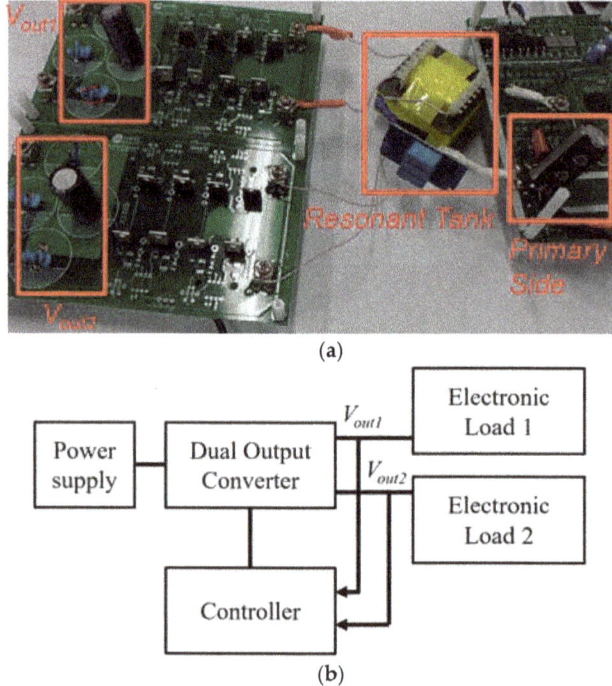

Figure 11. Experimental condition: (**a**) Prototype converter, (**b**) diagram of experimental setup.

The conventional cross regulation method has high output voltage error (6.1% for V_{o1} and 9% for V_{o2}) since its performance is not enough to compensate the output voltage error at the worst load condition. Figure 12 shows experimental waveforms of a prototype dual-output LLC resonant converter using the PFM-APWM hybrid control method at the worst load condition. The proposed control method shows much smaller output voltage error (0.25% for V_{o1} and 0.3% for V_{o2}) than that of the conventional method. It verifies the validity of the proposed design methodology with tight output voltage regulation at the worst load condition. The small voltage error of the hybrid control method might be composed of analog-to-digital conversion (ADC) and measurement errors.

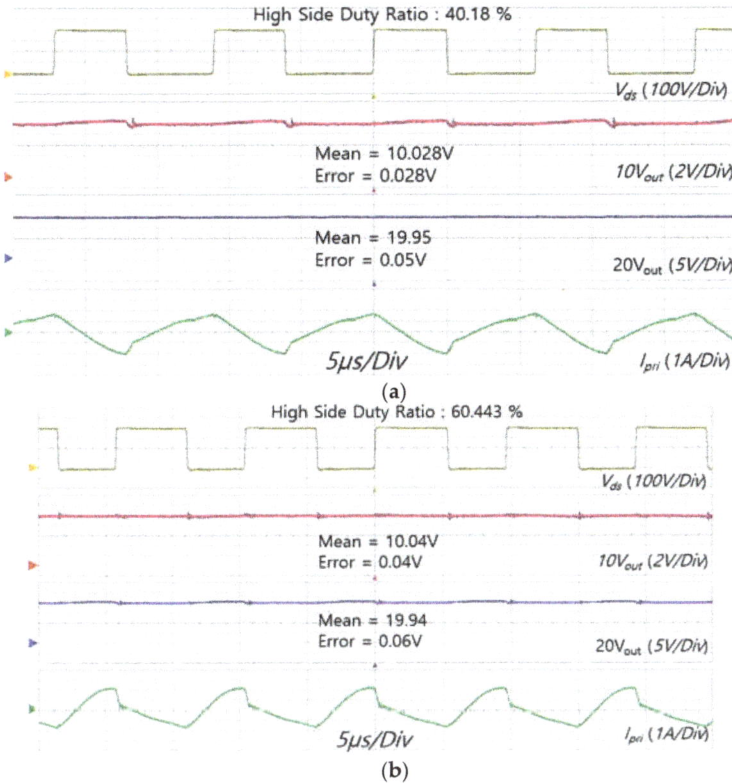

Figure 12. Experimental waveforms of the output voltage regulation with hybrid control algorithm: (**a**) $I_{out1} = 1$ A and $I_{out2} = 7$ A, (**b**) $I_{out1} = 6$ A and $I_{out2} = 1$ A.

Figure 13 shows the voltage error in cross regulation according to the load variation. In Figure 12, large load difference causes bigger errors of the output voltages. All the detail measured values of the simulation and experimental results are shown in Table 2 which shows the performance comparison of the cross regulation in the output voltages according to the control method. When two output powers are similar, the conventional and hybrid control algorithm has good output voltage regulation performance. When two output powers are significantly different, the conventional control algorithm has poor output voltage regulation. However, the hybrid control algorithm with the proposed design methodology can tightly regulate the output voltage for the entire load difference. The hybrid method shows more than 20 times less voltage regulation errors than those of the conventional simple cross regulation method at the worst condition.

Figure 14 shows the step load response of the prototype dual-output LLC resonant converter using the PFM-APWM hybrid control method which regulates the output voltages under load variations. There are no oscillations and disturbances in the converter operating waveforms. The results indirectly verify operating stability of the dual-output converter controlled by the hybrid control algorithm. Figure 15 shows the step load response of the conventional cross regulation method. The conventional regulation method has poor output voltage regulation according to the load condition. At the worst condition, the output voltage regulation performance is 24 times poorer compared with the hybrid control algorithm. The poor output voltage regulation is shown in Table 2. Figure 16 shows the power conversion efficiency of the prototype converter according to the load variations which show the load change in one output channel when the other is set to a fixed value. The proposed algorithm reduces the offset current on the magnetizing inductance compared with the case of the conventional method. However, the APWM operation induces a higher turn-off loss than that of the conventional method. Figure 16b shows the loss analysis according to the control method. Figure 16a shows the power conversion efficiency between the conventional and proposed methods. The power conversion efficiency is similar between the two control methods.

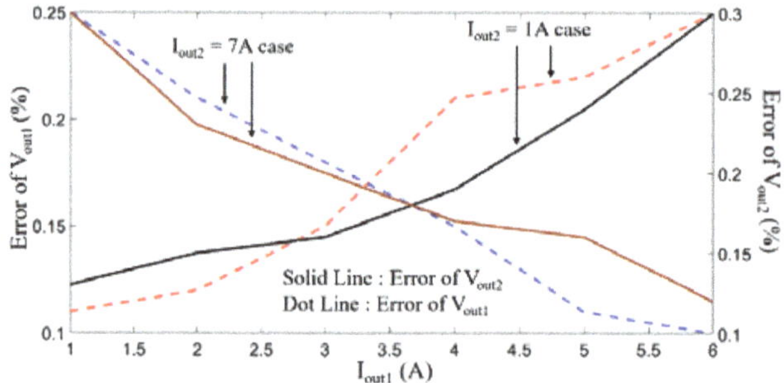

Figure 13. Error of output voltages according to the load variation.

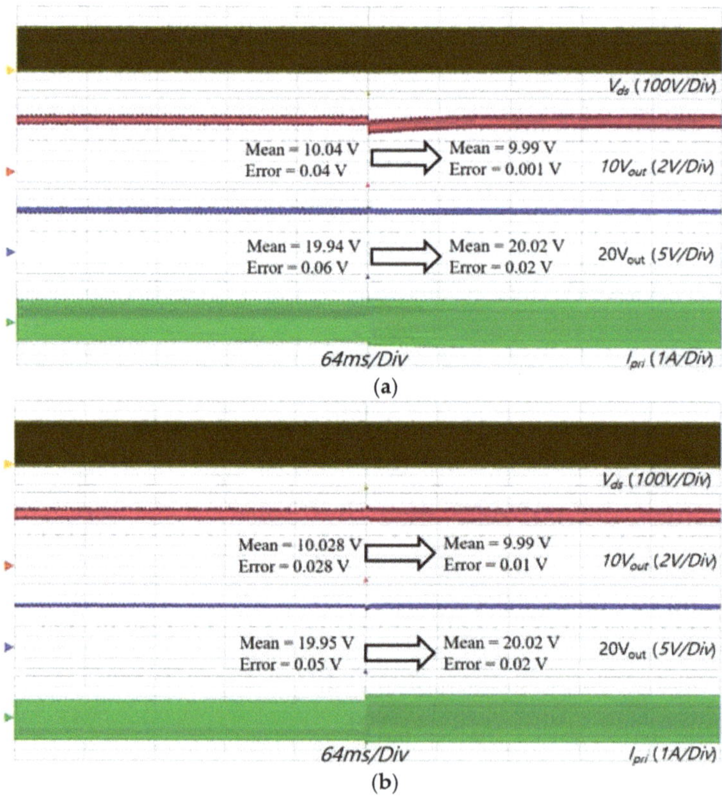

Figure 14. Step load responses of the dual-output converter with hybrid control algorithm: (**a**) I_{out1} = 6 A and I_{out2} changed from 1 A to 7 A, (**b**) I_{out1} changed from 1 A to 6 A and I_{out2} = 7 A.

Table 2. Performance comparison by experimental verifications.

	Conventional Cross Regulation	**Proposed Cross Regulation**
k_w Duty	$k_{w1} = 1$ and $k_{w2} = 1$ $D = 1$	$k_{w1} = 1$ and $k_{w2} = 1$ $0.35 < D < 0.65$
Case 1 Error	$I_{out1} = 1$ A and $I_{out2} = 7$ A $V_{out1} = 5\%$ and $V_{out2} = 9\%$	$V_{out1} = 0.25\%$ and $V_{out2} = 0.3\%$
Case 2 Error	$I_{out1} = 6$ A and $I_{out2} = 1$ A $V_{out1} = 6.1\%$ and $V_{out2} = 8.8\%$	$V_{out1} = 0.3\%$ and $V_{out2} = 0.3\%$
Case 3 Error	$I_{out1} = 1$ A and $I_{out2} = 1$ A $V_{out1} = 0.25\%$ and $V_{out2} = 0.34\%$	$V_{out1} = 0.12\%$ and $V_{out2} = 0.18\%$
Case 4 Error	$I_{out1} = 6$ A and $I_{out2} = 7$ A $V_{out1} = 0.41\%$ and $V_{out2} = 0.63\%$	$V_{out1} = 0.1\%$ and $V_{out2} = 0.12\%$

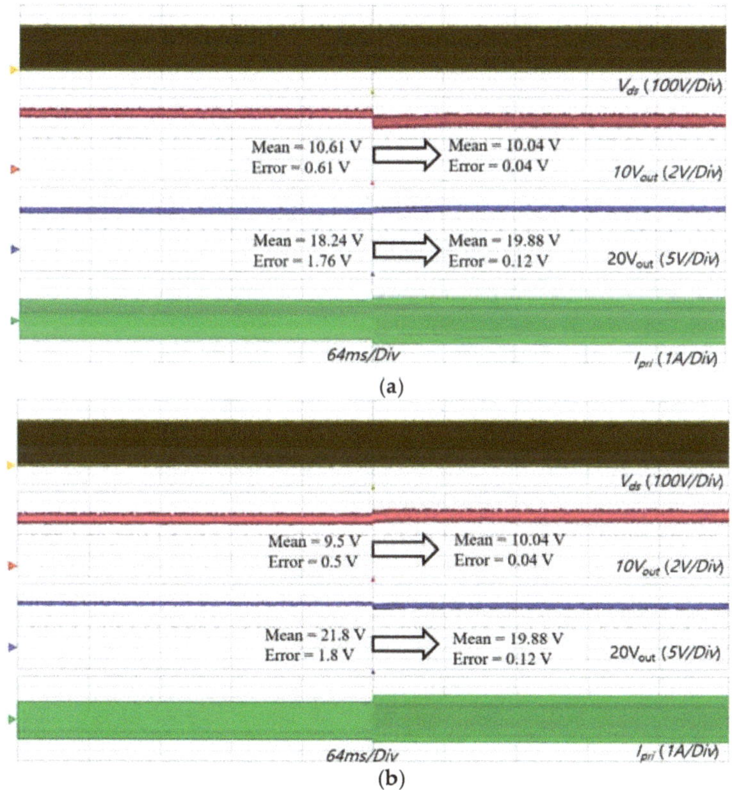

Figure 15. Step load responses of the dual-output converter with conventional cross regulation method: (a) $I_{out1} = 6$ A and I_{out2} changed from 1 A to 7 A, (b) I_{out1} changed from 1 A to 6 A and $I_{out2} = 7$ A.

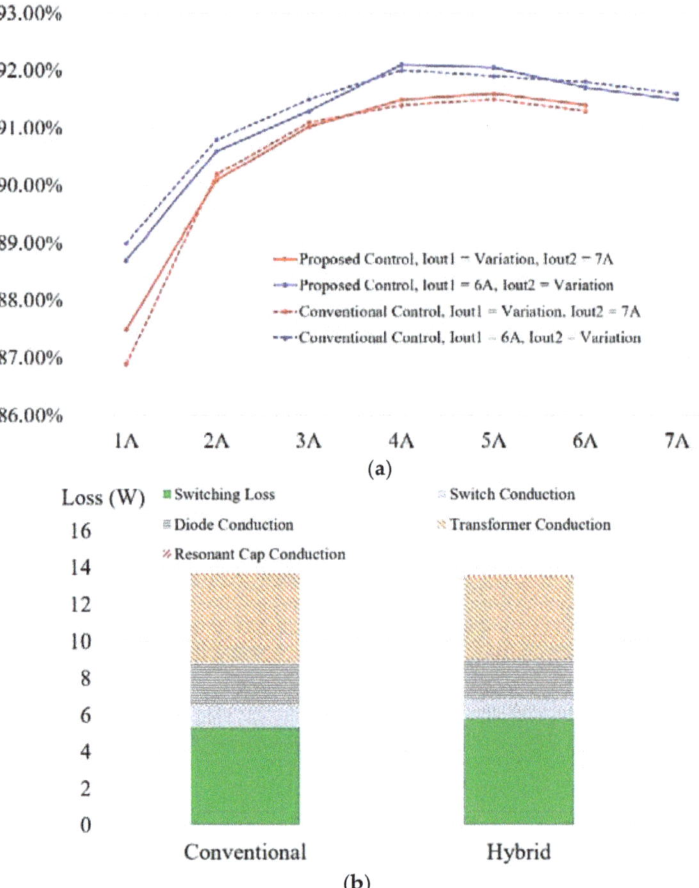

Figure 16. Comparison of power conversion efficiency and loss according to the control algorithm: (**a**) Power conversion efficiency, (**b**) loss analysis.

5. Conclusions

In this paper, the design methodology of a dual-output LLC resonant converter using the PFM-APWM hybrid control method is proposed to obtain tight output voltage regulation and ZVS capability for the entire load conditions. Through the analysis of the operational principle, the magnetizing inductance is designed to obtain ZVS capability. The resonant impedance is designed to implement the tight output voltage regulation for the entire load condition. The simulation and experimental results using the 190 W prototype converter verify the validity of the proposed design methodology, and the hybrid control algorithm. Without any power conversion efficiency degradation, the PFM-APWM hybrid control algorithm reduces the output voltage error to 24.4 times and 30 times smaller than those of the conventional cross regulation method at the worst load condition.

Author Contributions: Conceptualization, H.P., M.K., and J.J.; methodology, H.P.; software, H.P.; validation, H.P., M.K. and J.J.; formal analysis, H.P.; investigation, H.P.; resources, J.J.; data curation, M.K.; writing—original draft preparation, H.P.; writing—review and editing, M.K., J.J., H.K.; visualization, H.P.; supervision, J.J. and H.K.; project administration, J.J.; funding acquisition, J.J.

Funding: This work was supported by the Samsung Research Funding Center of Samsung Electronics under Project Number SRFC-TA1403-51.

Conflicts of Interest: The authors declare no conflict of interest.

References

1. Xi, Y.; Jain, P.K. A forward converter topology with independently and precisely regulated multiple outputs. *IEEE Trans. Power Electron.* **2003**, *18*, 648–658.
2. Chen, Q.; Lee, F.C.; Jovanovic, M.M. Analysis and design of weighted voltage mode control for a multiple-output forward converter. In Proceedings of the Eighth Annual Applied Power Electronics Conference and Exposition, San Diego, CA, USA, 7–11 March 1995; pp. 449–455.
3. Ji, C.; Smith, M.; Smedley, K.M.; King, K. Cross regulation in flyback converters: Analytic model and solution. *IEEE Trans. Power Electron.* **2001**, *16*, 231–239.
4. Kim, H.S.; Jung, J.H.; Baek, J.W.; Kim, H.J. Analysis and Design of a Multi output Converter Using Asymmetrical PWM Half-Bridge Flyback Converter Employing a Parallel-Series Transformer. *IEEE Trans. Ind. Electron.* **2013**, *60*, 3115–3125. [CrossRef]
5. Kim, J.W.; Ha, J.I. Dual voltage regulation of single switch flyback converter using variable switching frequency. In Proceedings of the IEEE Applied Power Electronics Conference and Exposition (APEC), Long Beach, CA, USA, 20–24 March 2016; pp. 1398–1402.
6. Lo, Y.K.; Yen, S.C.; Song, T.H. Analysis and Design of a Double-Output Series- Resonant DC-DC Converter. *IEEE Trans. Power Electron.* **2007**, *22*, 952–959. [CrossRef]
7. Ayachit, A.; Reatti, A.; Kazimierczuk, M.K. Magnetizing inductance of multiple- output flyback dc/dc convertor for discontinuous-conduction mode. *IET. Power Electron.* **2017**, *10*, 451–461. [CrossRef]
8. Jung, J.H.; Ahmed, S. Flyback converter with novel active clamp control and secondary side post regulator for low standby power consumption under high efficiency operation. *IET Power Electron.* **2011**, *4*, 1058–1067. [CrossRef]
9. Wang, X.; Tian, F.; Batarseh, I. High efficiency parallel post regulator for wide range input dc-dc converter. *IEEE Trans. Power Electron.* **2008**, *23*, 852–858. [CrossRef]
10. Kim, J.K.; Lee, J.B.; Moon, G.W. Zero-voltage switching multioutput flyback converter with integrated auxiliary buck converter. *IEEE Trans. Power Electron.* **2014**, *29*, 3001–3010. [CrossRef]
11. Kim, E.H.; Lee, J.J.; Kwon, J.M.; Choi, W.Y.; Kwon, B.H. Asymmetrical pwm half-bridge converter with independently regulated multiple outputs. *IEE Proc. Electr. Power Appl.* **2006**, *153*, 14–22. [CrossRef]
12. Su, B.; Wen, H.; Zhang, J.; Lu, Z. A soft-switching post regulator for multi-outputs dual forward dc/dc converter with tight output voltage regulation. *IET Power Electron.* **2013**, *6*, 1067–1077. [CrossRef]
13. Park, S.G.; Ryu, S.H.; Cho, K.S.; Lee, B.K. An improved single switched post regulator for multiple output isolated converters. In Proceedings of the IEEE Power Electronics Conference (ECCE), Seoul, Korea, 1–5 June 2015; pp. 2403–2408.
14. Kim, J.K.; Choi, S.W.; Kim, C.E.; Moon, G.W. A new standby structure using multi output full-bridge converter integrating flyback converter. *IEEE Trans. Ind. Electron.* **2011**, *58*, 4763–4767. [CrossRef]
15. Hwu, K.I.; Ziang, W.Z.; Kim, C.E.; Moon, G.W. Time-sharing pwm control scheme for isolated multi-output dcdc converter. *Electron. Lett.* **2015**, *51*, 1446–1447. [CrossRef]
16. Nami, A.; Zare, F.; Ghosh, A.; Blaabjerg, F. Multi-output dc-dc converters based on diode-clamped converters configuration: Topology and control strategy. *IET Power Electron.* **2010**, *3*, 197–208. [CrossRef]
17. Ray, O.; Josyula, A.P.; Mishra, S.; Joshi, A. Integrated dual-output converter. *IEEE Trans. Ind. Electron.* **2015**, *62*, 371–382. [CrossRef]
18. Trevisan, D.; Mattavelli, P.; Tenti, P. Digital control of single inductor multiple output step-down dc-dc converters in ccm. *IEEE Trans. Ind. Electron.* **2008**, *55*, 3476–3483. [CrossRef]
19. Wang, B.; Kanamarlapudi, V.R.K.; Xian, L.; Peng, K.T.T.X.; So, P.L. Model predictive voltage control for single-inductor multiple-output dcdc converter with reduced cross regulation. *IEEE Trans. Ind. Electron.* **2016**, *63*, 4187–4197. [CrossRef]
20. Chen, Y.T.; Shih, F.Y. New multi-output switching converters with mosfet-rectifier post regulators. *IEEE Trans. Ind. Electron.* **1998**, *45*, 609–616. [CrossRef]
21. Park, H.; Jung, J. PWM and PFM Hybrid Control Method for LLC Resonant Converters in High Switching Frequency Operation. *IEEE Trans. Ind. Electron.* **2017**, *64*, 253–263. [CrossRef]

22. Park, H.; Jung, J. Power Stage and Feedback Loop Design for LLC Resonant Converter in High-Switching-Frequency Operation. *IEEE Trans. Power Electron.* **2017**, *32*, 7770–7782. [CrossRef]
23. Choi, Y.-J.; Cha, H.-R.; Jung, S.-M.; Kim, R.-Y. An Integrated Current-Voltage Compensator Design Method for Stable Constant Voltage and Current Source Operation of LLC Resonant Converters. *Energies* **2018**, *11*, 1325. [CrossRef]
24. Liu, Y.-C.; Chen, C.; Chen, K.-D.; Syu, Y.-L.; Tsai, M.-C. High-Frequency LLC Resonant Converter with GaN Devices and Integrated Magnetics. *Energies* **2019**, *12*, 1781. [CrossRef]
25. Demirel, I.; Erkmen, B. A very low-profile dual output llc resonant converter for lcd/led tv applications. *IEEE Trans. Power Electron.* **2014**, *29*, 3514–3524. [CrossRef]
26. Roes, M.G.L.; Duarte, J.L.; Hendrix, M.A.M. Disturbance observer-based control of a dual-output llc converter for solid-state lighting applications. *IEEE Trans. Power Electron.* **2011**, *26*, 2018–2027. [CrossRef]
27. Cho, S.H.; Kim, C.S.; Han, S.K. High-efficiency and low cost tightly regulated dual-output llc resonant converter. *IEEE Trans. Ind. Electron.* **2012**, *59*, 2982–2991. [CrossRef]
28. Hang, L.; Wang, S.; Gu, Y.; Yao, W.; Lu, Z. High cross-regulation multioutput llc series resonant converter with magamp postregulator. *IEEE Trans. Ind. Electron.* **2011**, *58*, 3905–3913. [CrossRef]
29. Wu, H.; Wan, K.; Sun, K.; Xing, Y. A high step-down multiple output converter with wide input voltage range based on quasi two stage architecture and dual output llc resonant converter. *IEEE Trans. Power Electron.* **2015**, *30*, 1793–1796. [CrossRef]
30. Elferich, R.; Duerbaum, T. A new load resonant dual-output converter. In Proceedings of the IEEE Power Electronics Conference (ECCE), Cairns, Australia, 23–27 June 2002; pp. 1319–1324.
31. Park, H.; Kim, M.; Jung, J. Tightly regulated dual-output half-bridge converter using PFM-APWM hybrid control method. In Proceedings of the IEEE Applied Power Electronics Conference and Exposition (APEC), Tampa, FL, USA, 26–30 March 2017; pp. 2022–2026.
32. Yang, B.; Lee, F.C.; Zhang, A.J.; Huang, G. LLC resonant converter for front end DC/DC conversion. In Proceedings of the IEEE Applied Power Electronics Conference and Exposition (APEC), Dallas, TX, USA, 10–14 March 2002; pp. 1108–1112.
33. Jung, J.H. Feed-forward compensator of operating frequency for apwm hb flyback converter. *IEEE Trans. Power Electron.* **2012**, *27*, 211–223. [CrossRef]

Article

Current Mode Control of a Series LC Converter Supporting Constant Current, Constant Voltage (CCCV)

Michael Heidinger *, **Qihao Xia, Christoph Simon, Fabian Denk, Santiago Eizaguirre, Rainer Kling and Wolfgang Heering**

Karlsruhe Institute for Technology (KIT)—Light Technology Institute, Engesser Straße 13, 76131 Karlsruhe, Germany
* Correspondence: michael.heidinger@kit.edu; Tel.: +49-721-608-47852

Received: 22 May 2019; Accepted: 11 July 2019; Published: 20 July 2019

Abstract: This paper introduces a control algorithm for soft-switching series LC converters. The conventional voltage-to-voltage controller is split into a master and a slave controller. The master controller implements constant current, constant voltage (CCCV) control, required for demanding applications, for example, lithium battery charging or laboratory power supplies. It defines the set-current for the open-loop current slave controller, which generates the pulse width modulation (PWM) parameters. The power supply achieves fast large-signal responses, e.g., from 5 V to 24 V, where 95% of the target value is reached in less than 400 µs. The design is evaluated extensively in simulation and on a prototype. A match between simulation and measurement is achieved.

Keywords: control; current mode control; voltage control; transfer function; power converter; soft-switching converter; battery charging

---

1. Introduction

By the use of soft-switching converters, highly efficient DC/DC converters can be built. One possible topology, a series LC (SLC) converter is shown in Figure 1. The topology is similar to a series resonant converter, but it operates in a non-resonant push–pull mode [1]. In contrast to a dual active half-bridge converter, the two secondary side active output switches are replaced with diodes [2].

A detailed time domain analysis for calculating the SLC output current, operated above the LC resonance frequency, was published recently [1]. Current literature proposes a voltage-to-voltage transfer function [3,4]. We split the voltage-to-voltage converter in a cascaded structure [2,5] for enhanced performance. A master controller sets the SLC output current, while a slave open-loop transfer function controls the switching period, duty cycle, and pulse-skipping. By the use of current mode control, one pole is eliminated in the control loop [6]. The master voltage controller supports constant current, constant voltage (CCCV) operation, which is required, e.g., for battery charging [7] or laboratory power supplies.

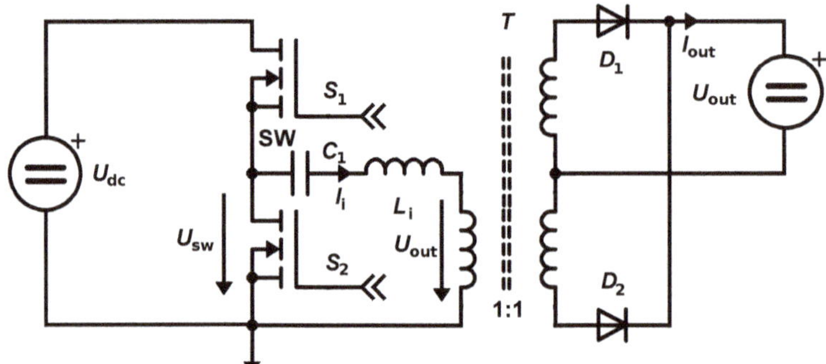

Figure 1. The schematic of the series LC converter is identical to the series resonant converter. However, the resonant capacitor C_1 is chosen large and acts as a DC blocking capacitor. The converter is operated far above its resonance frequency.

2. State-of-the-art

State-of-the-art soft-switching converters, e.g., series resonant converters (SRC) or LLCs, are modeled using a voltage transfer function [3,4]. This function can be derived, for example, by the first harmonic approximation. Current research modeled resonant converters operated far above the resonant frequency, so-called series LC converters, by a voltage-to-current transfer function [1]. Hence, the idea of current mode control suggests itself.

Current mode control has been used in flyback converters for a very long time [8]. Thereby, the loop is split into a master voltage controller and a slave current mode controller [5,8]. This approach has already been shown for dual active half-bridge [2] and resonant converters [6]. Thereby, current mode control has a multitude of advantages [6]. State-of-the-art closed loop current mode control uses a low pass filter [6], resulting in a reduced bandwidth. This work uses the open-loop voltage-to-current transfer function, which has a higher bandwidth, to further enhance performance [1].

Series resonant converters are typically controlled by switching frequency only [4]. However, pulse skipping and duty cycle modulation have also been presented for current limiting [9]. Research also demonstrated linear open-loop feed-forward control [10], whereas this work uses non-linear open-loop feed-forward control, based on [1].

3. Fundamentals

Equation (1) formulates the SLC converter output current I_{cc} as a function of the input voltage U_{dc} and output voltage U_{out} based on the control parameters duty cycle D and switching period t_p [1]. By adding pulse skipping, where p_o and p_c represent the pulse skipping parameter defined in Figure 2, a very large output current range is achieved.

$$I_{cc} = \frac{p_o}{p_c} \frac{D(1-D)U_{dc}^2 - U_{out}^2}{4L_i U_{dc}} t_p \tag{1}$$

As the input voltage and also the output voltage are monitored in (1), this equation allows a very high rejection ratio [1]. The output voltage is U_{out}, and the measured output voltage is referred to as U_{meas}. As a large input voltage ripple can be rejected, this allows for a reduction of the DC link capacitance C_{dc}. This enables the use of film capacitors instead of electrolytic capacitors, extending the estimated service life of the power supply.

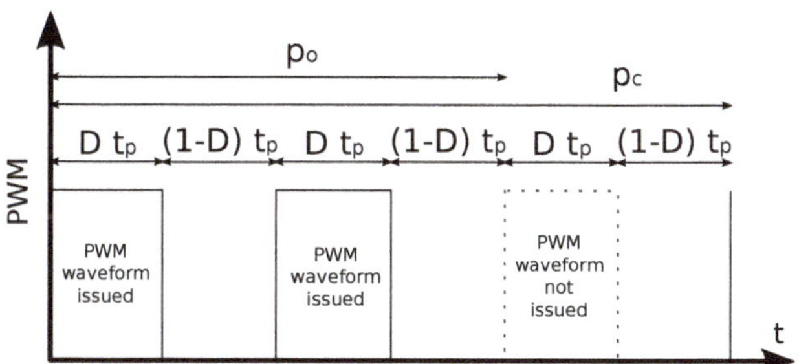

Figure 2. A PWM pulse skipping waveform is shown, where the period (t_p), duty cycle (D), and pulse skipping ontime ($p_\mathrm{o} = 2$) and pulse skipping period ($p_\mathrm{c} = 3$) are highlighted.

The proposed control diagram is shown in Figure 3. The control is based on four elements: {1} The master voltage controller sets the current to the slave current controller. {2} the slave current mode controller is an open-loop control transfer function based on (1). The current controller sets four parameters to the pulse width modulation (PWM) modulator {3}. The PWM modulator generates the PWM output waveform for the series LC converter {4}.

The controller is implemented on a digital signal processor (DSP), as (1) requires non-linear calculations. ADCs digitize the input voltage, output voltage, and output current for the CCCV control. The DSP integrates a PWM module, generating the gate signals for the half-bridge.

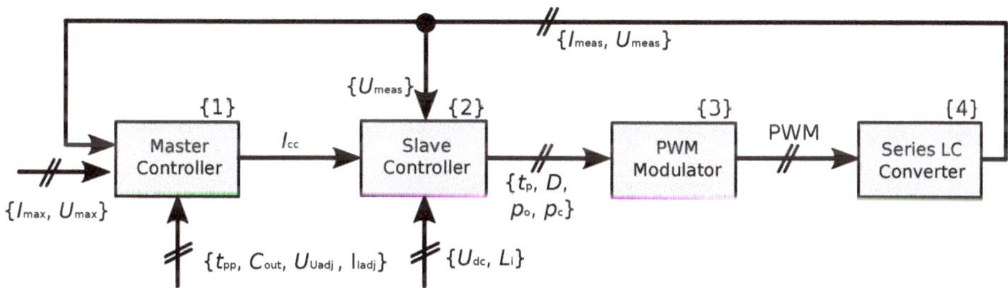

Figure 3. Proposed control diagram for the series LC converter. The converter is split into a master controller, implementing constant current, constant voltage (CCCV) control, and an open-loop slave controller. The MCUs pulse width modulator (PWM) is used to generate the SRCs gate signals.

4. Master Voltage Mode Controller

The master voltage mode controller is shown in detail in Figure 4. It consists out of two controllers: one to control the voltage and one to control the current. Both operate in parallel, and the minimum value is selected to control the set current I_{cc} for the constant current controller.

The sensed current I_{sense} should be filtered to prevent systems oscillation when a capacitive load is connected. For the prototype, presented in Section 7, a second order low pass filter with a cutoff frequency of 16 kHz is used. Both controllers are discussed in detail in the following subsections.

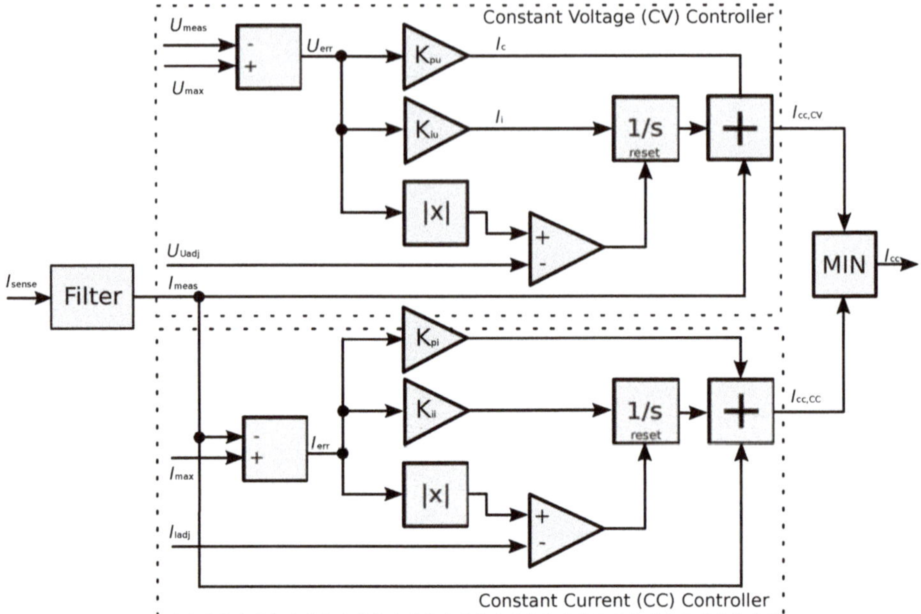

Figure 4. The master controller implements the CCCV functionality. The current and the voltage controllers operate in parallel, and the smaller set value is selected for the set current I_{cc}.

4.1. Constant Voltage Controller

The constant voltage controller limits the maximal output voltage to U_{max}. We design the voltage control loop in Figure 4 based on circuit analysis of the output capacitor C_{out}. The required set current $I_{\text{cc,CV}}$ is expressed in (2). The filtered current is designated I_{meas}.

$$I_{\text{cc,CV}} = I_{\text{meas}} + I_{\text{c}} + I_{\text{i}} \tag{2}$$

The equalization current ($I_{\text{c}} + I_{\text{i}}$) is calculated by a PI regulator. The proportional gain is chosen on the charge balance observation: we calculate the proportional equalizing current I_{c} as a function of the output charge.

$$Q = I_{\text{c}} \cdot t_{\text{pp}} = C_{\text{out}}(U_{\text{max}} - U_{\text{meas}}) \tag{3}$$

$$I_{\text{c}} = \frac{C_{\text{out}}(U_{\text{max}} - U_{\text{meas}})}{t_{\text{pp}}} \tag{4}$$

The time constant t_{pp} in (5) is chosen with respect to the maximal digital regulator control loop period. Stability was observed by using a factor of $1/4$ or less at a control loop frequency of $85.750\,\text{kHz}$. Hence, the P regulator gain can be formulated as:

$$K_{\text{pu}} \leq \frac{C_{\text{out}}}{4 \cdot t_{\text{pp}}} \tag{5}$$

The DC voltage accuracy is enhanced by increasing the proportional gain K_{pu}. Referring to (5), the output capacitance is proportional to the maximal proportional gain.

As shown in Figure 4, an I regulator with reset is used to achieve stationary accuracy of the control. It adjusts a typically small error. To reduce overshoot, it is only activated if the error is lower than an absolute minimal error. We name this minimal voltage U_{Uadj}.

4.2. Constant Current Controller

The constant current controller limits the output current to I_{max}. Previous research already demonstrated that the slave output current accuracy I_{cc} is better than 7% [1]. Therefore, the output current is directly forwarded to the limiter. To compensate for inaccuracies, an additional PI regulator is used. If the absolute error is larger than I_{Iadj}, the I regulator is reset to reduce overshoot for large signal responses.

4.3. Acoustic Noise

This design uses multilayer ceramic capacitors (MLCCs) as output capacitors. They may emit acoustic noise due to the capacitors piezoelectric dielectric. If the master P gain is chosen close to the critical gain, noise is emitted. The acoustic noise is reduced by lowering the P gain or choosing low noise MLCCs.

Our experiments concluded that the following control loop gain eliminated the noise at the cost of a slightly slower step response:

$$K_{pu} \leq \frac{C_{out}}{9 \cdot t_{PP}}. \tag{6}$$

5. Slave Current Controller

The slave current controller receives the set current I_{cc} from the master controller. It is responsible for selecting the appropriate modulation scheme; it chooses between adjusting the switching period, duty cycle, or pulse skipping, as shown in Figure 5.

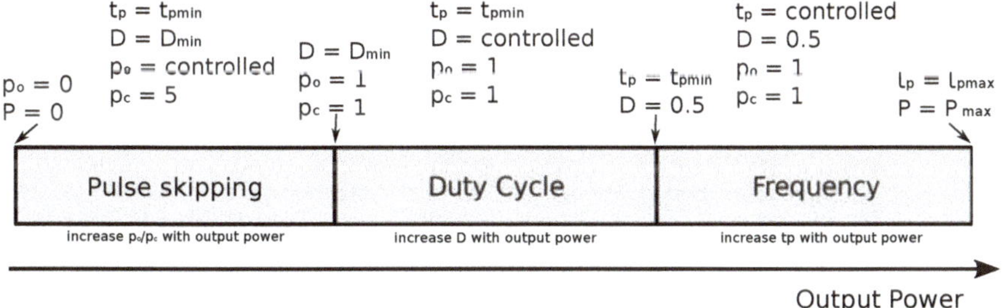

Figure 5. The slave current mode controller controls the frequency ($f_p = 1/t_p$) for high output power, the duty cycle (D) for medium output power, and uses pulse skipping (p_o/p_c) for low output power. Changing the modulation scheme maximizes the output current range.

For adjusting the output power, we propose the modulation strategy shown in Figure 5. Pulse skipping is used for the lowest possible output power. It is identified by missing PWM pulses. The number of pulses may range between zero and p_c. At medium output power, the duty cycle is in the range of $D = D_{min,abs}$ to $D = 0.5$, while the minimal switching period is used. This modulation is referred to as duty cycle modulation. The minimal duty cycle is a design parameter and is chosen to $D_{min,abs} = 0.2$.

At very high output power, the switching period is increased until the maximal allowed $t_{p,max}$ is reached. Frequency modulation uses a duty cycle of 0.5 and a variable switching frequency. In the case where $D_{max} < 0.5$, duty cycle ramp-up is used. It uses the minimal switching frequency while increasing D_{max} by steps of ΔD. Duty cycle ramp-up reduces stress on capacitor C_1 and prevents overcurrent.

The implementation of the algorithm for determining the appropriate modulation scheme is visualized in Figure 6 and is discussed in detail next.

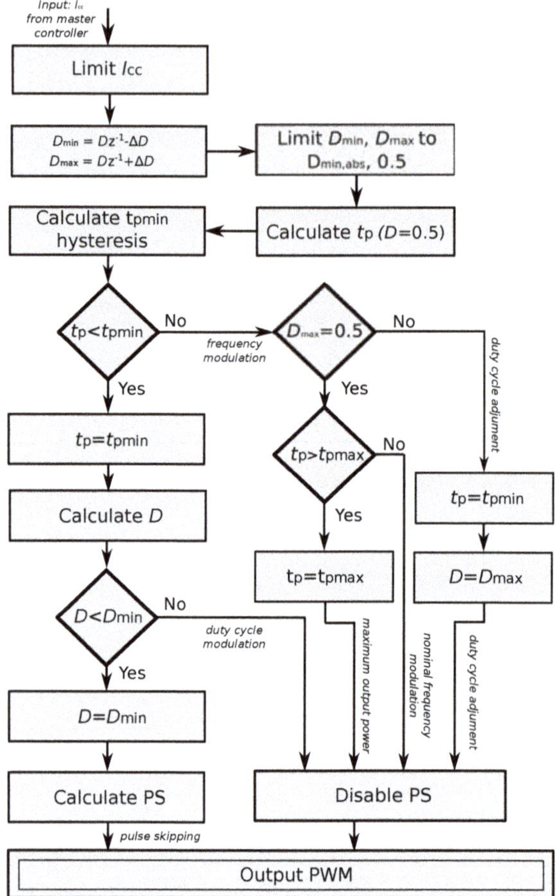

Figure 6. Slave current mode controller algorithm.

5.1. Initial Calculus

The algorithm shown in Figure 6 first limits I_{cc} to a positive value. Second, the minimal and maximal duty cycles are calculated, which are next limited. Third, the period t_p is determined using (7), which is based on (1), using a duty cycle of $D = 0.5$.

$$t_p = \frac{16 L_i U_{dc} I_{cc}}{U_{dc}^2 - 4 U_{meas}^2}. \tag{7}$$

Next, the minimum period $t_{p,min}$ is calculated, which is a constant value, with an additional hysteresis. In the simulation and experiments, no hysteresis was utilized. The minimum switching frequency $t_{p,min}$ is typically chosen in such a manner that soft-switching of the half-bridge is still achieved.

5.2. Frequency Modulation

If the period t_p is larger than $t_{p,min}$, switching frequency modulation is used. To prevent unintentional false-triggering due to capacitor C_1 displacement current, the minimal period $t_{p,min}$ is used during duty cycle adjustment. A duty cycle ramp-up is in progress when $D_{max} < 0.5$.

The maximum switching period is chosen based on (8). The constant k should be chosen in the range of 0.5 to 0.7, depending on the design goals. A lower k results in less slave controller error, while a higher k allows a higher output current.

$$t_{p,\text{max}} = k\pi\sqrt{L_i C_1} \tag{8}$$

5.3. Duty Cycle Modulation

If the period t_p is smaller than $t_{p,\text{min}}$, duty cycle modulation is used. The duty cycle is determined by (1). As a quadratic equation has two results, one has to be chosen. To limit the voltage stress on C_1, the smaller result is used. This results in the following equation:

$$D = \frac{U_{\text{dc}}\, t_{p,\text{min}} - \sqrt{\left(U_{\text{dc}}^2 - 4U_{\text{meas}}^2\right) t_{p,\text{min}}^2 - 16 I_{cc}\, L_i\, U_{\text{dc}}\, t_{p,\text{min}}}}{2U_{\text{dc}}\, t_{p,\text{min}}}. \tag{9}$$

5.4. Pulse Skipping

If the calculated duty cycle is smaller than the minimum duty cycle D_{min}, pulse skipping is used. The pulse modulation is calculated according to the following formula, which is converted to the number of on-pulses (p_o) and total number of periods (p_c). An example for a PWM waveform with pulse skipping is given in Figure 2.

$$I_{cc} = \frac{p_o}{p_c} \frac{D_{\text{min}}(1 - D_{\text{min}}) U_{\text{dc}}^2 - U_{\text{meas}}^2}{4 L_i U_{\text{dc}}} t_p \tag{10}$$

Currently, a fixed pulse skipping period p_c is used. However, the Farey method could also be used to determine a more accurate ratio [11]. If $p_o < 0.5$, the PWM output is disabled. Thereby, very low pulse counts are achieved. To prevent acoustic noise by pulse skipping, the pulse skipping frequency should be larger than the maximal audible frequency $f_a = 20\,\text{kHz}$:

$$\frac{1}{p_c\, t_{p,\text{min}}} > f_a. \tag{11}$$

Pulse skipping introduces a significant amount of output ripple. Currently, output ripple can be reduced by increasing the output capacitor C_{out}. To further reduce output ripple, an appropriate LC filter and its impact on the output response could be investigated.

5.5. Voltage Stress on C_1

The voltage U_c on the offset capacitor C_1 is calculated using the following equation [1]:

$$U_c = D U_{\text{dc}}. \tag{12}$$

To limit the voltage slope stress on the offset capacitor C_1 and slow down its aging, the duty cycle is only changed slowly. Currently, a value of $\Delta D = 0.02$ per iteration is used to limit the stress on the DC blocking capacitor C_1.

5.6. Input Voltage Range

The minimal input voltage must be at least twice the output voltage when a duty cycle of 50% is applied. Furthermore, a transformer ratio of 1:1 and no output current is assumed. The following equation is derived based on (1).

$$U_{\text{out,max}} = 0.5\, U_{\text{dc}} \tag{13}$$

To determine the minimum DC link voltage for a given output voltage U_{out} and an output current I_{out}, (1) is solved for $U_{\text{dc,min}}$.

$$U_{\text{dc,min}} = \frac{2\sqrt{U_{\text{out}}^2 t_{\text{p,max}}^2 + 16 I_{\text{cc}}^2 L_{\text{i}}^2} + 8 I_{\text{cc}} L_{\text{i}}}{t_{\text{p,max}}} \tag{14}$$

The minimum input voltage calculated by (14) is plotted for the converter as a function of output voltage in Figure 7. It shows that, the SLC converter allows a lower minimum input voltage when a reduced output voltage range is sufficient. Therefore, input and output voltage range may be traded off.

Equation (14) shows that the larger the maximum switching period $t_{\text{c,max}}$ is, the less influence the output current I_{out} has on the minimal input voltage $U_{\text{dc,min}}$. This is visualized in Figure 8. The minimal input voltage $U_{\text{dc,min}}$ is shown as a function of the maximal cycle time $t_{\text{p,max}}$ at an output voltage of $U_{\text{out}} = 25$ V. According to (8), the output current range is increased by choosing a larger capacitance C_1.

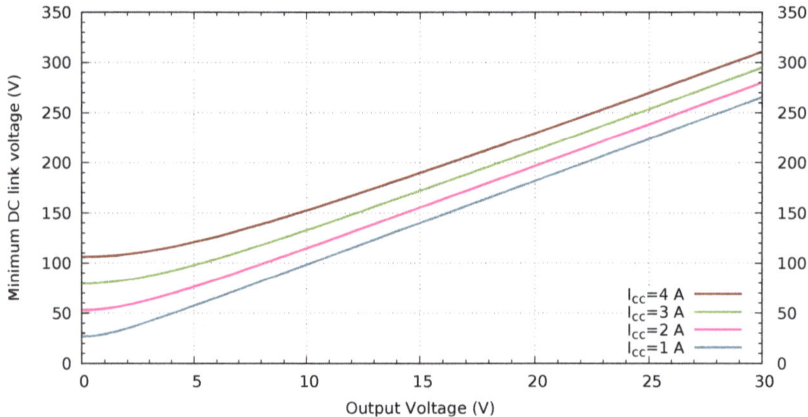

Figure 7. The minimum DC link voltage is plotted over the output voltage of the converter. The output current I_{cc} is shown as a parameter.

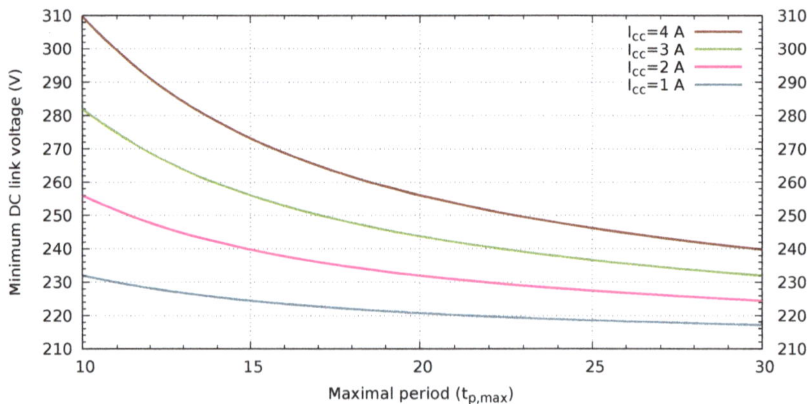

Figure 8. The minimum DC link voltage is plotted over the maximal allowable period for an output voltage $U_{\text{out}} = 25$ V. The output current I_{cc} is shown as a parameter.

6. Modulator

The DSP PWM unit generates the gate signals. It has four input parameters: The period t_p, the duty cycle D, the number of emitted pulses p_o, and the number of pulses per period p_c. An example is depicted in Figure 2. The period $t_p = \frac{1}{f_{sw}}$ is the inverse of the switching frequency, and the duty cycle states the ratio of the PWM high period. A switching cycle can be skipped by pulse skipping. The pulse skipping ontime p_o states how many PWM pulses are emitted during a pulse skipping period p_c.

7. Simulation and Experimental Results

The following section covers the simulation and measurement results for the CCCV converter.

7.1. Measurement Setup

To verify operation, the circuit was simulated with the software PLECS and tested in an experimental setup. The build converter prototype is shown in Figure 9. The test parameters are shown in Table 1, unless otherwise noted in the measurement description. For the simulations and experiments, a load resistor of $R_{load} = 10\,\Omega$ was connected to the output.

Four experiments were carried out on the prototype: {1} a constant voltage step response test, {2} a constant current step response test, {3} a load response test, and {4}, the CCCV step response. For each test setup, the corresponding output current and voltage were measured. In addition to each experiment, output voltage and current were simulated as well. The depicted duty cycle and switching period are extracted from simulation only. In the fifth simulation, the converter was operated at 230 V, 50 Hz AC, demonstrating its ability to reject a large input voltage DC link ripple.

Table 1. Test setup parameters and conditions.

Element/Parameter	Value
U_{in}	325 V
T_{ratio}	4.2:1
L_i	110 µH
C_1	470 nF
C_{out}	110 µF
$P_{out,max}$	62.5 W
K_{pu}	1.0
K_{iu}	857.5
U_{Uadj}	0.05 U_{set}
K_{pi}	20
K_{ii}	17,150
I_{Iadj}	0.05 I_{set}
ΔD	0.02
$D_{min,abs}$	0.2
$t_{p,min}$	5 µs
$t_{p,max}$	15.8 µs
k	0.7
p_c	5
$f_{control}$	85.750 kHz

Figure 9. The prototype is mounted in a DIN rail case measuring 85 mm by 65 mm. It avoids electrolytic capacitors and replaces them with film capacitors to achieve a longer service life.

7.2. Voltage Step Response

The constant voltage controller limits the allowable output voltage. To verify the constant voltage controller, the maximal output voltage U_{max} was increased in a step response at $t = 0$ from 5 V to 24 V; the step response is shown in Figure 10.

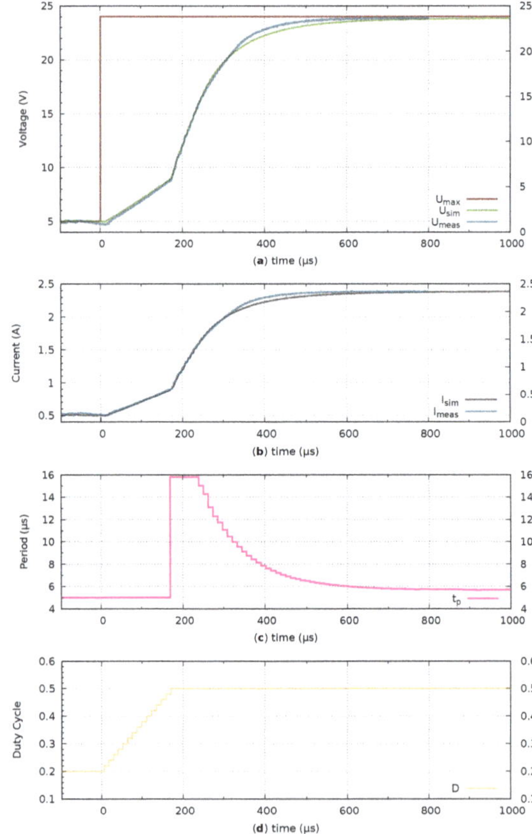

Figure 10. The constant voltage operation of the control was verified: the output voltage limit U_{max} was increased in a step response from 5 V to 24 V (**a**) at $t = 0$. The simulated and measured output voltages are shown in (**a**). The simulated and measured output current are shown in (**b**). The simulated switching period t_{p} is shown in (**c**), while the simulated duty cycle D is shown in (**d**).

Before the step at $t < 0$, the duty cycle D was limited to $D_{\min} = 0.2$ as pulse skipping was used. When the output power increased after $t = 0$, the duty cycle was increased during duty cycle ramp-up. The duty cycle D increased each control loop iteration by steps of $\Delta D = 0.02$ to $D = 0.5$. At $t \approx 180$ µs, the switching period was limited in frequency modulation to provide the maximal save output current while ensuring over-resonant operation. When the output voltage U_{meas} was about to reach the maximal voltage $U_{\max}$, the switching frequency was reduced. The converter remained in frequency modulation.

In Figure 10, the output voltage rose fast and reached 95% of the target output voltage in less than 400 µs. A transition from pulse skipping modulation at low load to switching frequency modulation at high load was demonstrated.

When the experiment's output voltage U_{meas} is compared to the simulated output voltage U_{sim} in Figure 10, a match is observed. The non-congruence between $t \approx 300$ µs and $t \approx 600$ µs arises due to the non-linear MLCC output capacitance.

7.3. Current Step Response

The constant current controller limits the maximum allowable output current. For verification, the maximal output current $I_{\max}$ was increased from 1 A to 2 A. The output current step response is shown in Figure 11. Before $t = 0$, the converter operated in pulse skipping mode. Pulse skipping generated a significant ripple on the output voltage, as the effective switching frequency is very low.

At $t = 0$, the output current I_{max} was increased from 1 A to 2 A, and the duty cycle was increased during duty cycle ramp-up from $D = 0.2$ to $D = 0.5$ in increments of $\Delta D = 0.02$ per control iteration, while using the minimal period t_p to prevent overcurrent triggering. To charge the output capacitor C_{out} fast, the slave controller first utilized frequency modulation, but switched back to duty cycle modulation at $t \approx 230$ µs as the output capacitor is almost charged.

Current control reached 95% of the output current target in $t \approx 300$ µs. At $t < 400$ µs, the fine adjustment by the PI regulator was complete. The converter did not overshoot on the output current.

When the measured output current from the experiment I_{meas} is compared to the simulated output current I_{sim} in Figure 11, a match is observed. The slight difference is due to the non-linear MLCC output capacitance.

7.4. Load Response

The load response monitors the output voltage change while the load is increased. For this experiment, shown in Figure 12, the output voltage was held constant at $U_{\max} = 5$ V, while an external current load was increased from $I_{\mathrm{meas}} = 0.5$ A to $I_{\mathrm{meas}} = 2$ A at $t = 0$. The output capacitor was chosen for this experiment to $C_{\mathrm{out}} = 160$ µF. Based on other system requirements, the control loop speed was reduced to 75 kHz for this experiment. To completely eliminate overshoots, the voltage PI controller was modified to $K_{\mathrm{pu}} = 0.9$ and $K_{\mathrm{iu}} = 343$. For reference, the simulated standard parameter response is shown in orange. It shows a simulated overshoot of less than 1%.

Before $t < 0$, the converter operated in pulse skipping mode, explaining the significant output ripple. At $t = 0$, the output current is I_{out} increased to 4 A. Therefore, the slave controller utilized duty cycle ramp-up, in which it increased the duty cycle by $\Delta D = 0.02$ per control cycle to $D = 0.5$. This limits capacitor stress on C_1 and prevents overcurrent. Because of that, the converter could not supply sufficient output power, hence the output voltage decreased. When the duty cycle adjustment was completed, at $t \approx 200$ µs, the controller switched to frequency modulation at $D = 0.5$. No overshoot of U_{meas} was observed. The controller then remained in frequency modulation. The simulation matches the measurement. It must be noted, that the presented load jump represents the worst case.

The load response may be improved by not using the minimal switching frequency during duty cycle adjustment. However, in that case, the overcurrent detection level has to be increased.

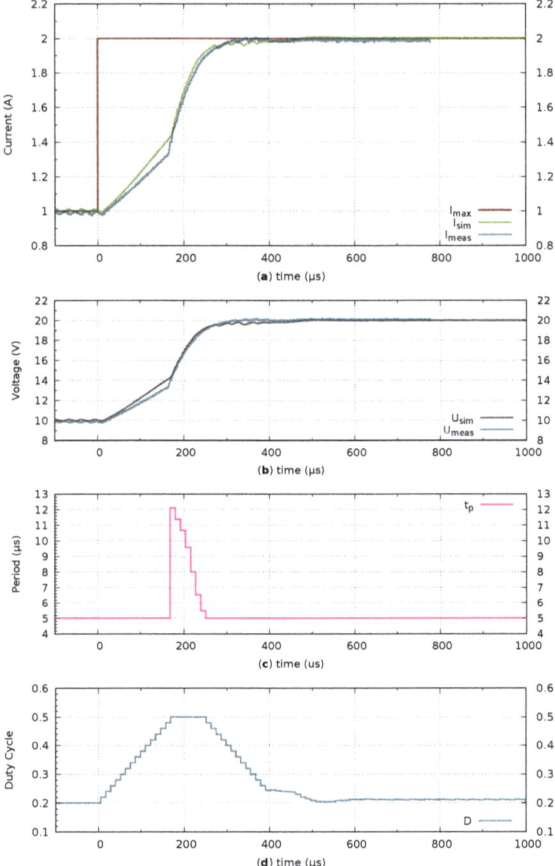

Figure 11. The constant current operation of the control was verified: the maximum output current I_{max} was increased in a step response from 1 A to 2 A (**a**) at $t = 0$. The simulated and measured output currents is shown also in (**a**). The simulated and measured output voltage is shown in (**b**), while the simulated switching period t_p is shown in (**c**), and the simulated duty cycle is shown in (**d**).

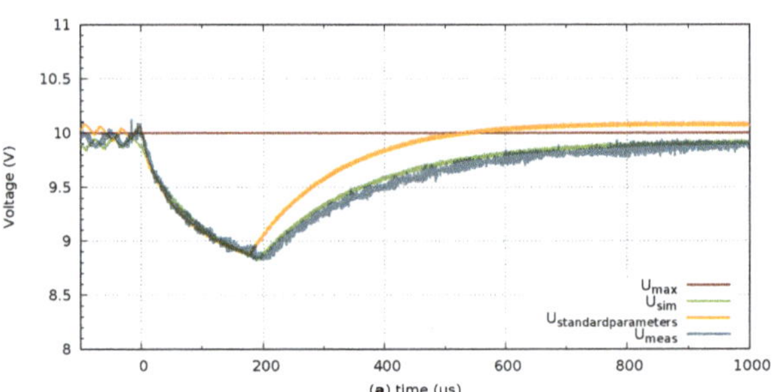

Figure 12. *Cont.*

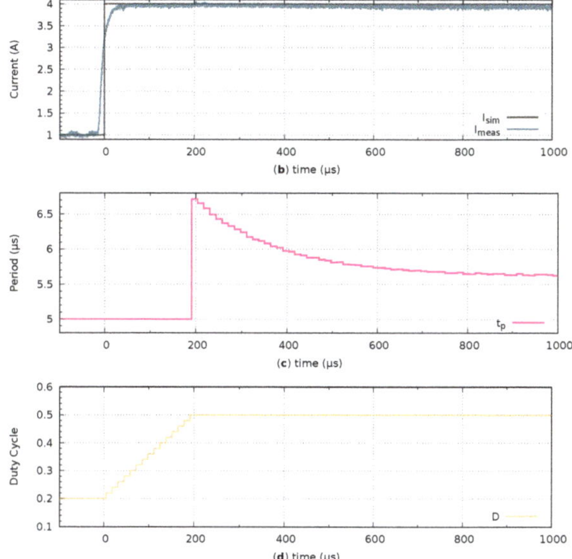

Figure 12. The load response of the power supply was measured: the external output current I_{out} was increased from 1 A to 4 A (**b**) at $t = 0$. The output voltage is measured in (**a**), where an output voltage drop is observed during duty cycle adjustment. The simulated and measured output currents are shown in (**b**), while the simulated switching period t_p is shown in (**c**), and the simulated duty cycle D is shown in (**d**).

7.5. CCCV Transition Step Response

In Figure 13, the constant current to constant voltage transition is measured. The effective output capacitance was determined to $C_{out} = 45\,\mu\text{F}$. The converter was first operated in constant current (CC) mode, and then the transition to constant voltage (CV) was demonstrated. The CCCV control was implemented by choosing the minimal value of two parallel controllers, as seen in Figure 4.

At $t < 0$, the converter operated in CC mode as the output current was limited to 2 A, while the slave controller operated in duty cycle modulation. The load $R_{load} = 10\ \Omega$ resulted in an output voltage of $U_{out} = 20$ V. The voltage limit was set to $U_{max} = 24$ V. At $t = 0$, the current limit I_{max} was increased from 2 A to 3 A. The master controller demanded additional output current, therefore the duty cycle was increased step wise during duty cycle ramp-up. This prevents false overcurrent triggering. Frequency modulation was used since $t \approx 170\ \mu\text{s}$, when the duty cycle ramp-up was completed, allowing a significantly increased output current and resulting in a faster output voltage rise. When the converter was about to reach its output voltage limit, the switching period was reduced. The converter continued to operate in frequency modulation. The output voltage rose from 20 V within 400 μs to the output voltage limit of $U_{max} = 24$ V. No output voltage overshoot was observed. A match between simulation and measurement is shown.

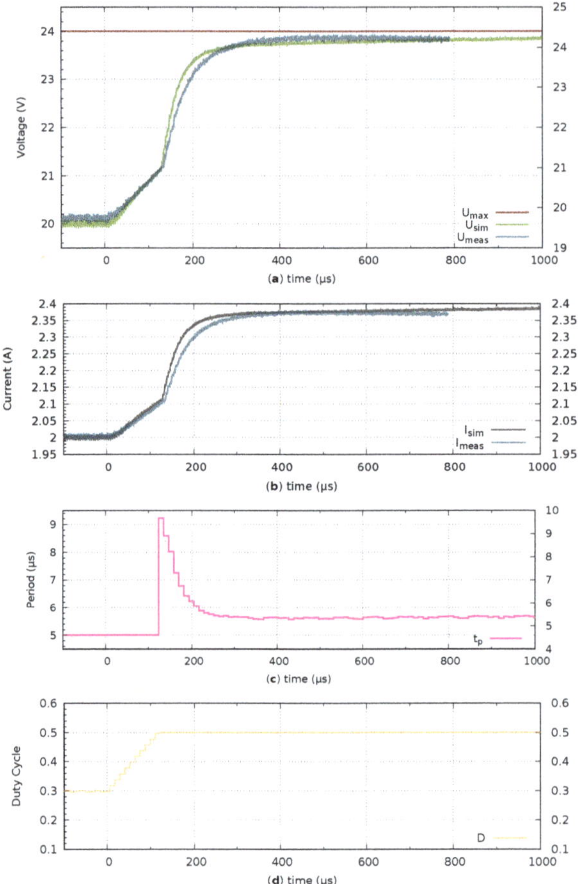

Figure 13. The constant current constant voltage operation of the control was verified: the maximal output current I_{max} was increased at $t = 0$ in a step response from 2 A to 3 A in (**b**) to demonstrate the CCCV behavior. The output voltage (**a**) limit was held constant at 24 V. Furthermore, the measured and simulated output voltage are shown. The simulated switching period t_p is shown in (**c**), while the simulated duty cycle D is shown in (**d**).

7.6. AC Input Voltage Range

To simulate the AC ripple rejection on the DC link, the converter was extended by a full bridge rectifier and a DC link capacitor $C_{in} = 30\,\mu\text{F}$ was used. The AC input voltage had a frequency of 50 Hz at an input voltage of $U_{ac,rms} = 230$ V. The output voltage was set to 25 V. The converter was loaded with a resistor of $10\,\Omega$. The voltage gain was chosen to $K_{pu} = 3$. The output voltage and input voltage are depicted in Figure 14 over half a typical line period. It can be seen that an input voltage range from 270 V to 325 V was accepted. Referring to Figure 14, the maximum switching period limit $t_{p,max}$ was not reached. Hence, a higher ripple is possible.

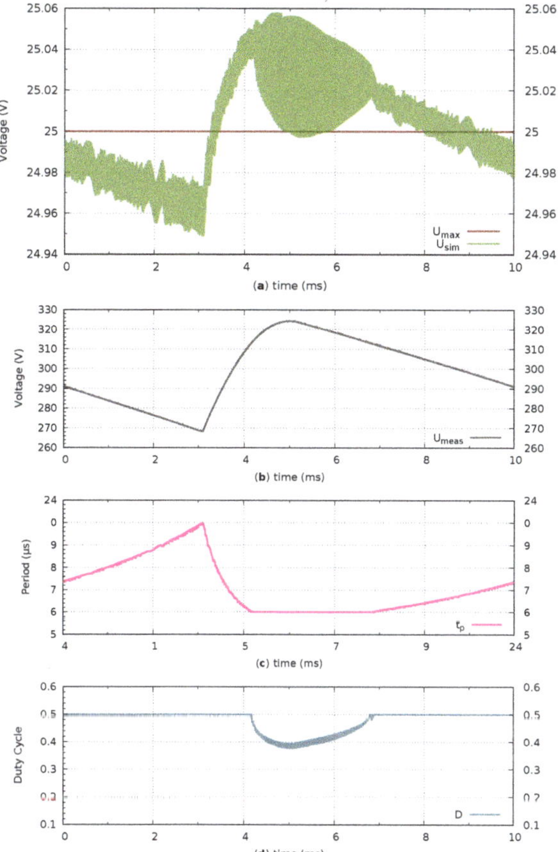

Figure 14. The operation of the converter at 50 Hz AC is simulated: The simulated output voltage U_{sim} is shown in (**a**). The simulated DC link voltage is shown also in (**b**), while the simulated switching period t_{p} is shown in (**c**), and the simulated duty cycle D is shown in (**d**).

7.7. DC Link Ripple Rejection

The input voltage range is extended by switching between the different modulation schemes. The ability to cope with a large voltage ripple allows for a smaller DC link capacitance. This simplifies the construction of electrolytic-free power supplies by replacing these by film capacitors.

The ripple gain is calculated by (15). Its inverse is the so-called ripple attenuation. A ripple gain close to zero results in a higher attenuation A. The higher the attenuation A, the better the converter rejects the DC link ripple on the output.

$$g = \frac{1}{A} = \frac{\dfrac{U_{\text{output,p2pripple}}}{U_{\text{output,peak}}}}{\dfrac{U_{\text{input,p2pripple}}}{U_{\text{input,peak}}}} \tag{15}$$

The converter ripple gain is measured to $g = 0.02$ and $A = 50$ in Figure 14.

7.8. Loop Gain Analysis

The loop gain analysis was conducted in simulation at an output voltage of 24 V and an amplitude of 0.1 V in CV mode. In Figure 15, a bandwidth of 1 kHz is simulated in the low acoustic noise design using a gain of 1/9. In a loop gain optimized design, using a factor of 1/4, a bandwidth of 3 kHz was observed. The acoustic noise of the loop gain optimized design is seen as jitter in the magnitude.

The converter's open-loop gain in Figure 15 has a typical integrator characteristic and suggests the well-tempered operation of the converter.

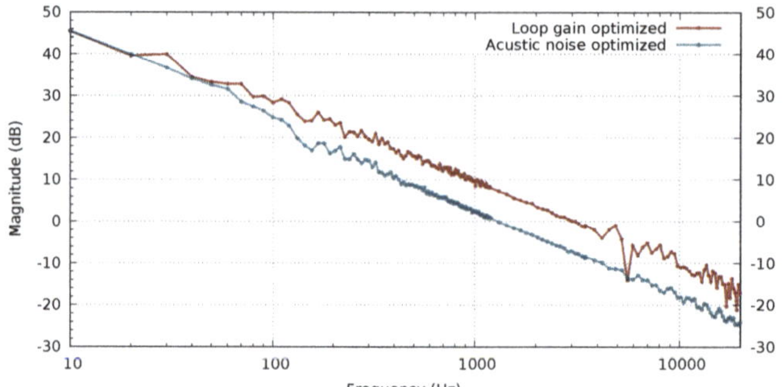

Figure 15. The simulated open-loop gain magnitude over frequency is shown for two design choices. The worst case bandwidth frequency is larger than 1 kHz.

8. Comparison with the State-of-the-art

In this section, the prototype is compared against the state-of-the-art converters.

8.1. Output Voltage Range

In comparison to conventional voltage-to-voltage converters, a larger output voltage range was achieved. The experiments and simulations demonstrated a large output voltage range from 5 V up to 25 V. This equals an output voltage range of 1:5. Currently, an output voltage range of 1:2 is considered large [9]. Hence, the presented output voltage range is 2.5 times larger compared to state-of-the-art soft-switching converters.

8.2. Input Voltage Range

The converter's input voltage range is shown in Figure 7. Experiments demonstrated an input voltage range from $U_{in,min} = 270$ V to $U_{in,max} = 325$ V. The output current of the converter is increased by choosing a larger resonance capacitor C_1 or by lowering the transformer ratio. Hence, a wide input voltage range is possible at the cost of a lower efficiency. Therefore, the SLC topology could be adopted for a large input voltage range.

8.3. DC Link Ripple Attenuation

The proposed converter shows, in Figure 14, a ripple gain of 0.02. Standard LLC converters have a ripple gain of 0.39 [12]. Hence, the presented control attenuates the DC link ripple approximately twenty times better compared to existing solutions.

8.4. Control Bandwidth

Conventional LLC converters have a typical bandwidth of 1.5 kHz [9] to 2 kHz [13] (p. 14) when connected to a restive load. Thus, referring to Figure 15, the presented control has a similar bandwidth compared to a typical LLC converter.

8.5. Overshoot

The cascaded current mode control changes the transient response characteristic of the converter from a state-of-the-art PT2 (second-order lag element) [14] to PT1 (first-order lag element). As one pole is eliminated in current mode control [6], no overshoot is observed in the prototype. State-of-the-art overshoot optimized designs show an typical overshoot of $\approx$5% [14].

8.6. Large Signal Step Response

The response time for CV, CC, and CCCV are very fast with $t_{resp,95\%} < 400$ µs. State-of-the-art converters feature a load responses time of 150 ms at an output voltage change from 40 V to 50 V [14].

9. Conclusions

The paper demonstrates fast and accurate control of a series LC converter in constant current, constant voltage mode and reliable transitions between those operational modes. Ninety-five percent of the output current and voltage target is reached in less than 400 µs during the step response test. The converter attenuates AC ripple on the DC link by a ratio of 1:50. The high attenuation is achieved due to the transfer function (1), canceling out input and output voltage variations. The high attenuation allows for a high DC link ripple. By this, high-capacitance electrolytic capacitors can be replaced with low-capacitance film capacitors without significantly increasing the converter size. Thereby, a significant increase in lifetime is expected.

In contrast to a resonant converter controlled in voltage mode, no overshoot was measured. This is achieved by splitting the controller into a master and a slave controller. The prototype demonstrated an output voltage range from 5 V to 25 V, which equals an extremely large dynamic range from 1:5.

The CCCV characteristic allows for the use of the converter for demanding applications, e.g., laboratory power supplies or lithium battery chargers. The CCCV control is implemented using a cascaded control loop: the master controller sets the SLC current and the open-loop slave controller controls the switching period, duty cycle, and pulse skipping. The converter is stable over the whole operation range, from no load to heavy load conditions.

10. Patents

The modulation scheme, which is based on Equation (1) is covered by a pending patent. Germany: DE 10 2018 216 749.4—"Verfahren zur Steuerung eines Serien-Resonanz-Wandlers".

Author Contributions: Conceptualization, M.H.; methodology, M.H.; simulation, M.H.; validation, Q.X.; formal analysis, W.H.; investigation, Q.X., M.H.; writing, original draft preparation, M.H.; writing, review and editing, C.S., F.D., S.E., W.H.; visualization, M.H.; supervision, R.K.; project administration, R.K.

Funding: This research received no external funding.

Conflicts of Interest: The authors declare no conflict of interest.

Abbreviations

The following abbreviations are used in this manuscript:

CC	Constant current
CCCV	Constant current, constant voltage
CV	Constant voltage
DSP	Digital signal processor
MLCC	Multilayer ceramic capacitor
PT1	First-order lag element
PT2	Second-order lag element
PWM	Pulse width modulation
SMPS	Switch mode power supply
SL	Series LC (inductor capacitor)
SLCC	Series LC (inductor capacitor) converter

References

1. Heidinger, M.; Simon, C.; Fabian, D.; Eziguerre, S.; Heering, W. Open Loop Current Control of a Series LC Converter by Duty Cycle and Frequency. *EPE J.* **2018**, submitted.
2. Tissières, M.; Askarian, I.; Pahlevani, M.; Rotzetta, A.; Knight, A.; Preda, I. A digital robust control scheme for dual Half-Bridge DC-DC converters. In Proceedings of the 2018 IEEE Applied Power Electronics Conference and Exposition (APEC), San Antonio, TX, USA, 4–8 March 2018; pp. 311–315. [CrossRef]
3. Liu, T.; Zhou, Z.; Xiong, A.; Zeng, J.; Ying, J. A Novel Precise Design Method for LLC Series Resonant Converter. In Proceedings of the INTELEC 06—Twenty-Eighth International Telecommunications Energy Conference, Providence, RI, USA, 10–14 September 2006; pp. 1–6. [CrossRef]
4. Patarau, T.; Petreus, D.; Duma, R.; Dobra, P. Comparison between analog and digital control of LLC converter. In Proceedings of the 2010 IEEE International Conference on Automation, Quality and Testing, Robotics (AQTR), Cluj-Napoca, Romania, 28–30 May 2010; pp. 1–6. [CrossRef]
5. Zhuang, M.; Atherton, D.P. Optimum cascade PID controller design for SISO systems. In Proceedings of the 1994 International Conference on Control—Control '94, Coventry, UK, 21–24 March 1994; Volume 1, pp. 606–611. [CrossRef]
6. Petkov, R.; Anguelov, G. Current mode control of frequency controlled resonant converters. In Proceedings of the INTELEC—Twentieth International Telecommunications Energy Conference (Cat. No.98CH36263), San Francisco, CA, USA, 4–8 October 1998; pp. 103–108. [CrossRef]
7. Aizpuru, I.; Iraola, U.; Canales, J.M.; Echeverria, M.; Gil, I. Passive balancing design for Li-ion battery packs based on single cell experimental tests for a CCCV charging mode. In Proceedings of the 2013 International Conference on Clean Electrical Power (ICCEP), Alghero, Italy, 11–13 June 2013; pp. 93–98. [CrossRef]
8. Kleebchampee, W.; Bunlaksananusorn, C. Modeling and Control Design of a Current-Mode Controlled Flyback Converter with Optocoupler Feedback. In Proceedings of the 2005 International Conference on Power Electronics and Drives Systems, Kuala Lumpur, Malaysia, 28 November–1 December 2005; pp. 787–792. [CrossRef]
9. Fariborz, M.; Marian, C.; Deepak, G.; Wilson, E. Control Strategies for Wide Output Voltage Range LLC Resonant DC–DC Converters in Battery Chargers. *IEEE Trans. Veh. Technol.* **2014**, *63*, 1117–1125. [CrossRef]
10. Brañas, C.; Azcondo, F.J.; Casanueva, R. Feedforward compensation of resonant converters with heavy ripple in the DC bus for LED lamp driver applications. In Proceedings of the 2013 IEEE 14th Workshop on Control and Modeling for Power Electronics (COMPEL), Salt Lake City, UT, USA, 23–26 June 2013; pp. 1–6. [CrossRef]
11. Graham, R.L.; Knuth, D.E.; Patashnik, O. *Concrete Mathematics: A Foundation for Computer Science*, 2nd ed.; Addison-Wesley: Boston, UK, 1989; ISBN 0-201-55802-5. Available online: https://www.csie.ntu.edu.tw/~r97002/temp/Concrete%20Mathematics%202e.pdf (accessed on 15 July 2019).
12. Texas Instruments Application Report: Ripple Rejection in AC/DC Systems Using LLC SLUA850. Available online: http://www.ti.com/lit/an/slua850/slua850.pdf (accessed on 15 July 2019).
13. Texas Instruments. 600-W, Isolated PFC Power Supply for AVR Amplifiers Based on the TAS5630 and TAS5631. Available online: http://www.ti.com/lit/ug/slou293c/slou293c.pdf (accessed on 15 July 2019).
14. Kurokawa, F.; Murata, K. A new quick response digital modified P-I-D control LLC resonant converter for DC power supply system. In Proceedings of the 2011 IEEE Ninth International Conference on Power Electronics and Drive Systems, Singapore, 5–8 December 2011; pp. 35–39. [CrossRef]

Article

PV Micro-Inverter Topology Using LLC Resonant Converter

Hiroki Watanabe [1], Jun-ichi Itoh [1,*], Naoki Koike [2] and Shinichiro Nagai [2]

[1]   Department of Electrical, Electronics and Information Engineering, Nagaoka University of Technology, Kamitomiokamachi, Nagaoka, Niigata 940-2137, Japan
[2]   Pony Electric Co., Ltd., Kitasaiwai, Yokohama, Kanagawa 220-0004, Japan
*   Correspondence: itoh@vos.nagaokaut.ac.jp; Tel.: +81-258-47-9561

Received: 11 July 2019; Accepted: 7 August 2019; Published: 13 August 2019

Abstract: In this paper, a DC–single-phase AC power converter with an LLC resonant converter is presented for a photovoltaic (PV) micro-inverter application. This application requires the leakage current suppression capability. Therefore, an isolated power converter is usually combined for DC/AC systems. The LLC resonant converter is the one of the isolated power converter topologies, and it has good performance for conversion efficiency with easy control. On the other hand, a double-line frequency power ripple has to be compensated for in order to improve the performance of the maximum power point tracking (MPPT). Therefore, a bulky electrolytic capacitor is usually necessary for the power converter. However, the electrolytic capacitor may limit the lifetime of the micro-inverter. This paper introduces the PV micro-inverter with a LLC resonant converter. In addition, the active power decoupling circuit is applied in order to compensate the double-line frequency power ripple by the small capacitor in order to eliminate the electrolytic capacitor. Finally, the transformer design is considered in order to reduce the transformer losses. As a result, the conversion efficiency of the LLC converter is improved by 1% when the litz wire has many strands.

Keywords: PV micro-inverter; LLC converter; high switching frequency; transformer loss

---

1. Introduction

Recently, photovoltaic systems (PVs) are actively researched as a sustainable power solution due to their attractive characteristics such as flexibility, high system efficiency, and low manufacturing cost. In order to enhance the utilization of PVs, the generative power optimization by MPPT is necessary. However, the decrease of the generative power due to the shade is a problem for PVs with the typical centralized power conversion system (PCS). In order to solve this problem, AC module systems using micro-inverters are attractive in comparison with large capacity inverters [1–4]. This is because the micro-inverters optimize the generation power at each PV module. Note that the micro-inverters consist of the DC/AC converter and the isolated DC/DC converter because the leakage current due to the stray capacitance of the PV panel should be avoided for safety reasons and for easy installation. In this case, a flyback converter or push-pull converter is often chosen because these circuits have some advantages, e.g., a low number of components.

On the other hand, the LLC resonant converter is also one choice for the isolated DC/DC converter, because it has good performance to reduce the switching losses utilizing Zero Voltage Switching (ZVS) and Zero Current Switching (ZCS) by series resonance [4–9]. In particular, the parasitic capacitor is discharged before being turned on by the negative current. On the other hand, each drain current approximately becomes zero when each switch is turned-off, owing to the sinusoidal resonance current. That means the turn-on and the turn-off switching losses are drastically reduced by the ZVS and ZVS operation. In addition, the volume of the transformer is drastically reduced owing to the high switching frequency operation being same as the push-pull converter or flyback converter. Note that in the single-phase AC grid systems, a double-line frequency ripple occurs in the input current due to the single-phase AC grid, and the current ripple decays the performance of the MPPT because the PV generative power becomes unstable. Therefore, the bulky electrolytic capacitor is usually installed on the DC side. However, the bulky electrolytic capacitor may limit the lifetime of the power converter because the micro-inverter has to operate on the high environmental temperature condition although the lifetime of the electrolytic capacitor depends on the Arrhenius law. In order to solve this problem, the active power decoupling topologies have been studied to compensate the current ripple [10,11]. These topologies utilize the small capacitor such as a firm or ceramic capacitor for the power decoupling. Thus, an electrolytic capacitor is not necessary.

In this paper, a PV micro-inverter using an LLC converter is presented. In addition, the active power decoupling circuit based on the boost converter is combined with the micro-inverter in order to eliminate the bulky electrolytic capacitor in order to improve the reliability. Finally, the conversion efficiency of the LLC resonant converter is evaluated using two type litz wire in order to reduce the skin effect due to the high switching frequency operation. The contribution of this paper is that the influence of the number of strands to the conversion efficiency is evaluated in order to improve the conversion efficiency of the high frequency power converter. A lot of papers have only considered the transformer loss or the inductor loss [12], and the relationship between the number of strands and the conversion efficiency is not clarified.

2. Circuit Configuration

Figure 1 shows the PV power generative system. Typically, the centralized inverter, as shown in Figure 1a, is employed, and the generative DC power is translated to single-phase AC. In this system, each generative power of the PV modules is regulated by the single inverter with a large power rating. On the other hand, the micro-inverter as shown in Figure 1b is integrated to the PV modules, and each power converter is connected to the single-phase grid directly. Therefore, it is possible to optimize the generative power at each of the PV modules.

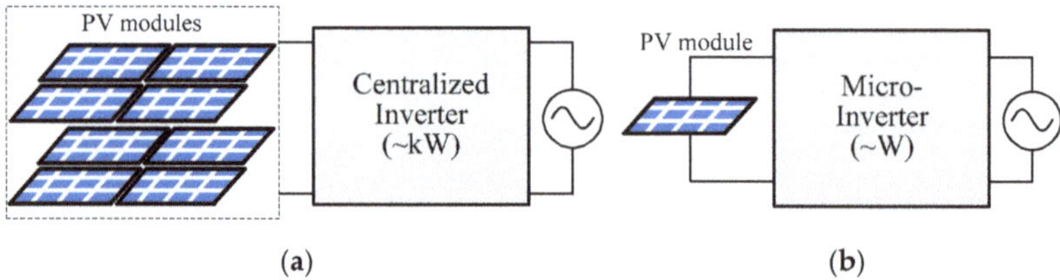

Figure 1. Configuration of PV power generative system. (**a**) Centralized inverter and (**b**) micro-inverter.

On the other hand, the micro-inverters require a long lifetime and the PV panels, likewise, are ideally maintenance-free. However, the electrolytic capacitor may decay over the lifetime of the converter. In particular, bulky electrolytic capacitor is necessary on the single-phase AC system in order to compensate the double-line frequency power ripple. In order to solve this problem, the micro-inverter with the active power decoupling circuit is considered in this paper.

Note that the micro-inverter systems are the distributed power supply, and each power converter transfers the generative power to the single-phase AC grid independently. Typically the all-converter operation is regulated by the Power Line Connection (PLC), and it is easy to get some information such as the power generation status. In this system, the micro-inverters are operated independently, and each power converters does not interfere with each other by the circuit operation.

Figure 2 shows the circuit configuration of the DC to single-phase AC converter which consists of LLC resonant converter and an active power decoupling circuit. The active power decoupling circuit compensates the double-line frequency power ripple due to the single-phase grid by the small capacitor [13]. The LLC resonant converter transfers the input power to the secondary side. After that, the boost-up of the DC link voltage and the power decoupling operation are achieved by the active power decoupling circuit. Finally, the grid connection is achieved by the Current Source Inverter (CSI). Note that the reverse-blocking diode is not necessary on CSI because the free-wheeling mode is implemented on the active power decoupling circuit.

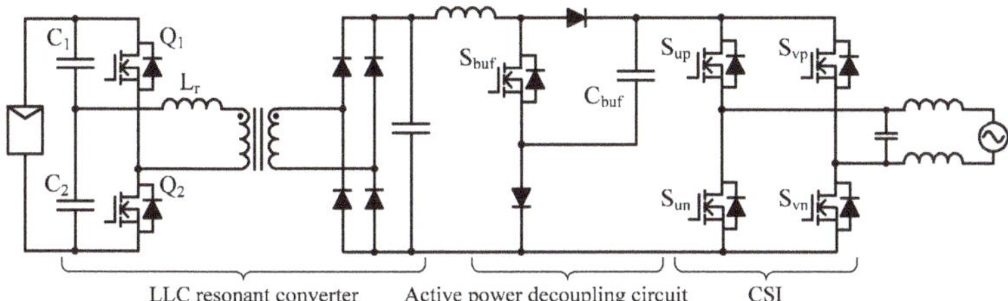

Figure 2. Circuit configuration of DC to Single-phase AC converter using LLC resonant converter.

3. Operation Principle of Active Power Decoupling Circuit

Figure 3 shows the principle of the power decoupling between the DC and single-phase AC sides [14]. In this system, the sinusoidal inverter output current is generated by the power converter in order to transfer the power to the single-phase AC grid. In this case, the single-phase grid voltage and the inverter output current are expressed as

$$v_{out} = V_{acp} \sin \omega t \tag{1}$$

$$i_{out} = I_{acp} \sin \omega t \tag{2}$$

where V_{acp} is the peak grid voltage and I_{acp} is the peak inverter current of the single-phase AC grid. When both the output voltage and the current waveforms are sinusoidal, the instantaneous output power pout is expressed as

$$p_{out} = v_{out} i_{out} = \frac{V_{acp} I_{acp}}{2} (1 - \cos 2wt) \tag{3}$$

from (3), the power ripple that contains double-line frequency of the power grid, appears at DC link.

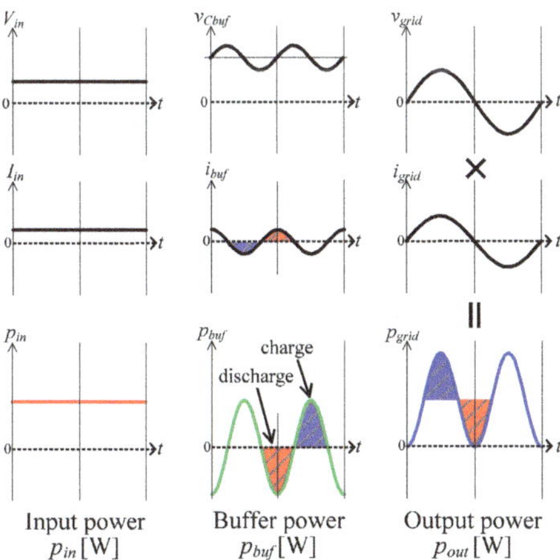

Figure 3. Principle of power decoupling between DC and single-phase AC.

In order to absorb the power ripple, the instantaneous power p_{buf} should be controlled by

$$p_{buf} = \frac{V_{acp}I_{acp}}{2} \cos 2\omega t \tag{4}$$

where the polarity of p_{buf} is defined as positive when the energy C_{fc} discharges. Due to the power decoupling capability, the input power is matched to the average output power of the single-phase AC. Therefore, the relationship between the input and output power is obtained as

$$p_{in} = \frac{1}{2}V_{acp}I_{acp} = V_{PV}I_{PV} \tag{5}$$

where V_{pv} is the PV input voltage and I_{pv} is the PV input current.

Figure 4 shows the operation mode of the active power decoupling circuit when the grid voltage is the positive period. This converter has the charge mode and discharge mode of the small buffer capacitor C_{buf} in order to regulate the capacitor voltage. In order to obtain the buffer power as per Equation (4), the capacitor voltage is actively fluctuated at the twice of the grid frequency such as Figure 3.

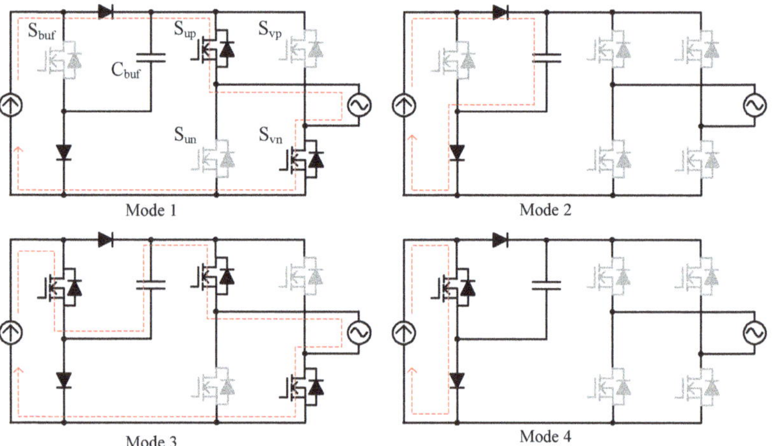

Figure 4. Operation mode of active power decoupling circuit.

The operation of the active power decoupling circuit cooperates with the CSI; this achieves both power decoupling and power transmission to the single-phase AC grid. In mode 1, the switch S_{buf} is turned on, and the input power is transferred to the single-phase AC grid. In mode 2, all the switches are turned off, and the buffer capacitor is charged. In mode 3, the buffer capacitor is discharged through the S_{buf} and CSI. Finally, mode 4 is the free-wheeling mode. The buffer capacitor voltage is regulated to adjust the duty command for modes 2 and 3 on the one period of the triangle carrier.

4. Operation Mode of LLC Resonant Converter

Figure 5 shows the operation mode, and the Figure 6 shows the key waveforms of the LLC resonant converter. In this circuit, the series resonance of C_1, C_2, and L_{res} is utilized for soft switching. The LLC converter regulates the input voltage of the active power decoupling circuit. The detail of the operation mode is as follows.

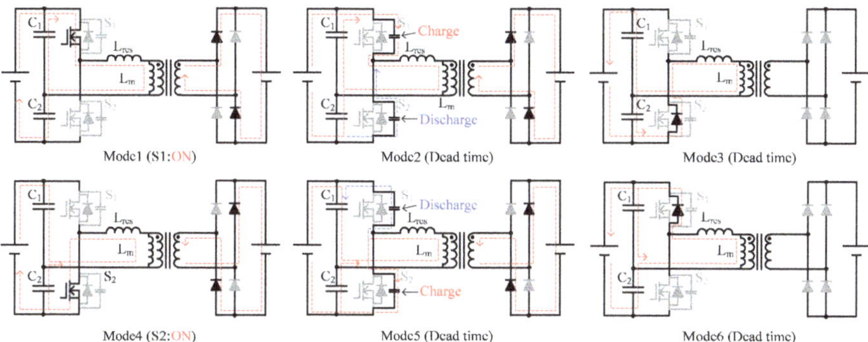

Figure 5. Operation mode of LLC resonant converter.

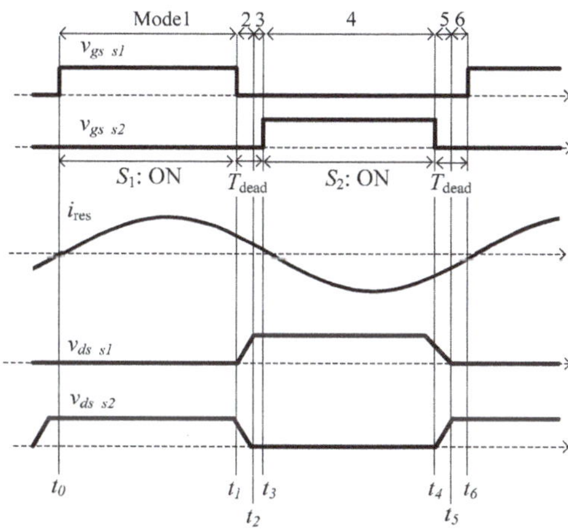

Figure 6. Key waveforms of LLC resonant converter.

Mode 1 [t_0-t_1]: S_1 is turned-on and half of the PV input voltage is applied to the transformer. In this mode, C_1, C_2, and L_{res} resonate, and the resonance current occurs.

Mode 2 [t_1-t_2]: S_1 is turned-off and the dead-time period is started. At this mode, S_1 achieves ZCS turned off. In mode 2, the parasitic capacitor C_{ds1} is quickly charged. On the other hand, the parasitic capacitor C_{ds2} is discharged owing to the negative resonant current. C_{ds2} is fully discharged in this mode, and mode 3 is started.

Mode 3 [t_2-t_3]: This mode is still dead-time, and the parasitic diode of S_2 conducts. Discharge of C_{ds2} should be completed during Mode2 and 3 for ZVS of S_2.

Mode 4 [t_3-t_4]: S_2 is turned on with ZVS when C_{ds2} is already discharged. In this mode, C_1, C_2, and L_{res} resonate as mode 1, and the resonant current occurs.

Mode 5 [t_4-t_5]: S_2 is turned off, and the dead-time period is started. At this mode, S_2 achieves ZCS turned-off. In mode5, the parasitic capacitor C_{ds1} is quickly charged. On the other hand, the parasitic capacitor C_{ds1} is discharged owing to the negative resonant current.

Mode 6 [t_5-t_6]: This mode is still dead-time, and the parasitic diode of S_1 conducts. Discharge of C_{ds1} should be completed during modes 5 and 6 for ZVS of S_1.

5. Experimental Result with Fundamental Operation

5.1. Active Power Decoupling Circuit and CSI

Table 1 shows the experimental parameters. Figure 7 shows the experimental result. Note that the output power is 300 W, the switching frequency is 20 kHz, the buffer capacitor is 44 μF, and the grid voltage is 200 V_{rms}.

Table 1. Experimental parameters.

Symbol	Quantity	Value
V_{in}	Input Voltage	90 V
P_{in}	Input Power	300 W
C_{buf}	Buffer Capacitor	44 μF
V_{ac}	Grid Voltage	200 V_{rms}
f_{ac}	Grid frequency	50 Hz
f_{sw}	Switching frequency	20 kHz

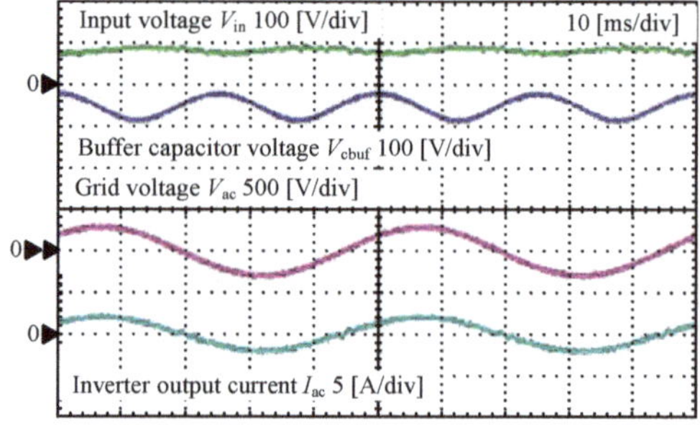

Figure 7. Experimental result of active power decoupling circuit and current source inverter (CSI).

According to Figure 7, the constant DC voltage and the sinusoidal inverter output current with the low THD are obtained. In addition, it is obtained that the buffer capacitor voltage is fluctuated by the double-line frequency by the power decoupling control. However, the low frequency component is still remaining on the input voltage. This is because the mismatch is occurring during the compensation value and the actual power ripple component. The compensation value for the power decoupling is obtained from the input power. However, the input power is included the conversion loss, which means the error between the compensation power and the actual output power occurs. This problem is solved by changing the calculation process of the compensation value to use the output power condition.

5.2. LLC Resonant Converter

Table 2 shows the experimental parameters and Figure 8 shows the experimental result. Note that the resonant frequency is designed to 205 kHz, and the switching frequency is set to below the resonant frequency as Table 2. In addition, the gate signals of S_1 and S_2 is provided by the 50% of duty signals from the controller. According to Figure 8, the sinusoidal resonant current is obtained by the series resonance.

Table 2. Experimental parameters.

Symbol	Quantity	Value
V_{in}	Input Voltage	40 V
V_{out}	Output Voltage	80 V
f_{sw}	Switching frequency	200 kHz
L_m	Magnetizing inductance	25 μH
L_{res}	Resonant inductance	400 nH
C_1, C_2	Resonant capacitance	750 nF
C_{dc}	DC capacitance	15 μF
C_j	Junction capacitance	500 pF
C_{oss}	Output capacitance	1500 pF

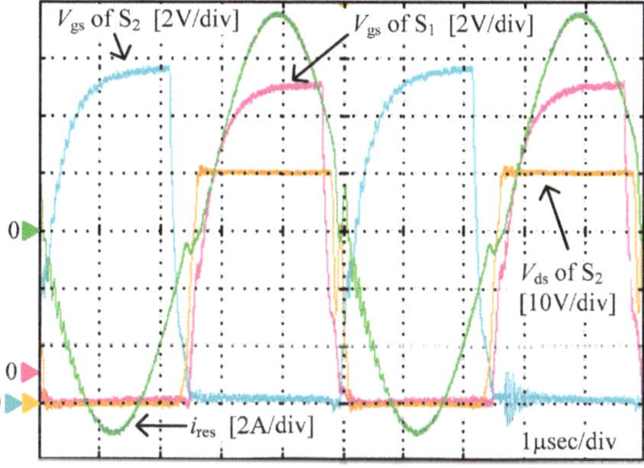

Figure 8. Switching waveforms.

Figure 9 shows the efficiency characteristics and the loss analysis result of the LLC resonant converter. According to Figure 9, 94.3% of the maximum efficiency is obtained when the output power is 100 W. In order to improve the conversion efficiency, the utilization of the synchronous rectifier is simple solution. On the other hand, the reduction of the transformer losses is important for the high switching frequency DC/DC converter because the copper loss becomes largely due to the skin effect and the proximity effect [15].

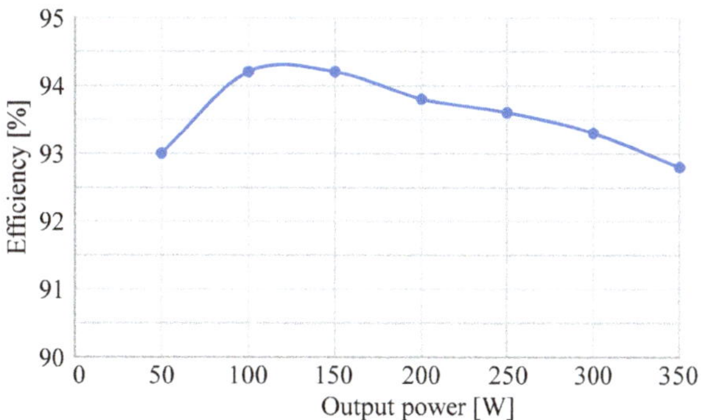

Figure 9. Efficiency curve.

6. Evaluation of Conversion Efficiency Using Different Litz Wire

In this section, the conversion efficiency of the LLC converter is evaluated in terms of a change the transformer. The high switching frequency operation increases the transformer loss including the copper loss due to the skin effect and the proximity effect, eddy loss, and hysteresis loss. Especially the skin effect and the proximity effect reduce the effective conduction area to surface of the wiring. As the result, the alternative wire resistance becomes large. In order to avoid this problem, the litz wire is usually used. In this paper, the two types of transformer are evaluated when the number of strands is changed.

Figure 10 shows the overview of the transformer, and Tables 3 and 4 show the parameter of the two transformers. Note that the air gap is not provided on this transformer in order to eliminate the leakage flax from the air gap, and the ferrite core with low eddy current loss is selected for the high switching frequency operation. The difference between two transformers is only number of strands, with transformer 2 having many strands on the wiring.

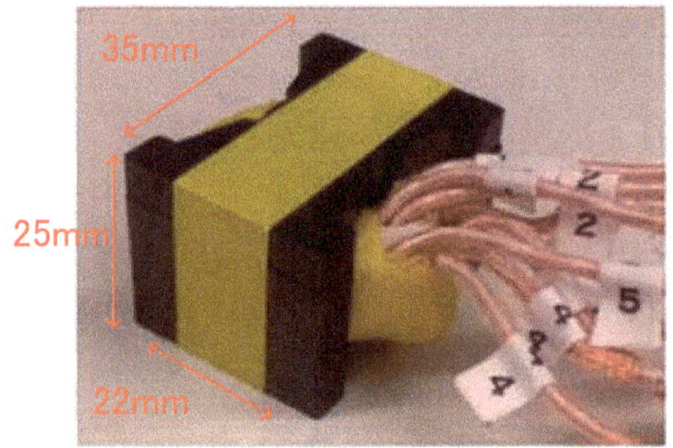

Figure 10. Overview of high frequency transformer.

Table 3. Parameters of transformer 1.

Symbol	Quantity	Value
N_1	Primary side number of turns	2
N_2	Secondary side number of turns	8
-	Litz wire specification	150/ϕ0.1
-	Current density	4 A/mm^2
-	Core material	Ferrite (PQ)
-	Air gap	0 mm
L_m	Magnetizing inductance	21.3 μH

Table 4. Parameters of transformer 2.

Symbol	Quantity	Value
N_1	Primary side number of turns	2
N_2	Secondary side number of turns	8
-	Litz wire specification	600/ϕ0.05
-	Current density	4 A/mm^2
-	Core material	Ferrite (PQ)
-	Air gap	0 mm
L_m	Magnetizing inductance	21.5 μH

Figure 11 shows the comparison of the conversion efficiency when the number of the strands is changed. Note that the orange and blue line means the approximate curve of each plot. The switching frequency is set to 200 kHz, and the same transformer core material is used. In this experiment, the conversion efficiency of the LLC converter with difference number and diameter of strands as shown in Table 3 is compared. According to Figure 11, it is obtained that the conversion efficiency is improved by 1% when the number of strand is increased on the wiring. This is because the increasing AC resistance due to the skin effect is suppressed, and the copper loss is reduced. Therefore, the litz wire with a lot of strands has a good performance in order to improve the conversion efficiency for the high switching frequency converter. Note that Figure 12 shows the comparison of the conversion efficiency when the switching frequency is changed. Note that the orange and blue lines represent the approximate curve of each plot. According to Figure 12, it is obtained that the improvement rate of the conversion efficiency is increased at 200 kHz. From this result, the number of strands should be chosen accordingly because it depends on the switching frequency.

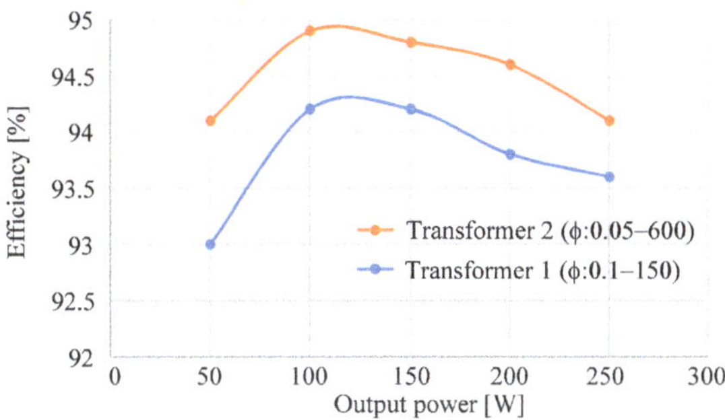

Figure 11. Comparison of conversion efficiency when number of strands is changed on wiring.

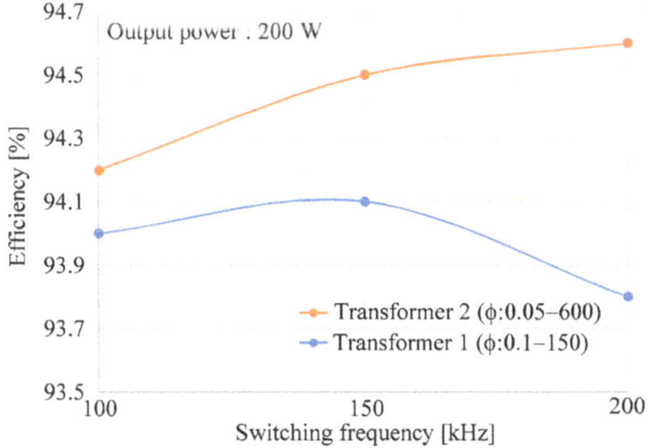

Figure 12. Comparison of conversion efficiency when switching frequency is changed.

7. Conclusions

In this paper, a DC–single-phase AC power converter with an LLC resonant converter is presented for a photovoltaic (PV) micro-inverter application. This converter consists of the LLC resonant converter, the active power decoupling circuit, and the CSI. In addition, the conversion efficiency of the LLC converter is evaluated when the number of strands is changed. Through the experimental result, it was obtained that the double-line frequency ripple is compensated by the small capacitor using an active power decoupling circuit. Finally, the conversion efficiency of the LLC converter is improved by 1% when the litz wire have a lot of strands.

In the future work, the optimal design of the high frequency transformer will be considered. In particular, the optimal number of the strand should be considered in order to minimize the transformer loss, and it depends on the switching frequency. Therefore, the relationship between the number of strands and the switching frequency will be analyzed. Finally, the optimal transformer design focusing on the minimum transformer losses will be demonstrated based on the loss analysis.

Author Contributions: Conceptualization, H.W. and J.-i.I.; methodology, H.W.; software, N.K.; validation, N.K., S.N. and J.-i.I.; formal analysis, H.W.; investigation, H.W.; resources, H.W.; data curation, H.W.; writing—original draft preparation, H.W.; writing—review and editing, H.W.; visualization, N.K.; supervision, J.-i.I.; project administration, J.-i.I.; funding acquisition, J.-i.I.

Funding: No external funding.

Acknowledgments: This study was supported by New Energy and Industrial Technology Development Organization (NEDO) of Japan.

Conflicts of Interest: The authors declare no conflicts of interest.

References

1. Lee, H.Y.; Yun, J.J. Quasi-Resonant Voltage Doubler with Snubber Capacitor for Boost Half-Bridge DC–DC Converter in Photovoltaic Micro-Inverter. *IEEE Trans. Power Electron.* **2019**, *34*, 8377–8388. [CrossRef]
2. Gautam, A.R.; Fulwani, D. Adaptive SMC for the Second-Order Harmonic Ripple Mitigation: A Solution for the Micro-Inverter Applications. *IEEE Trans. Power Electron.* **2019**, *34*, 8254–8264. [CrossRef]
3. Liao, C.Y.; Lin, W.S.; Chen, Y.M.; Chou, C.Y. A PV Micro-inverter with PV Current Decoupling Strategy. *IEEE Trans. Power Electron.* **2019**, *32*, 6544–6557. [CrossRef]
4. Leuenberger, D.; Biela, J. PV-Module-Integrated AC Inverters (AC Modules) with Subpanel MPP Tracking. *IEEE Trans. Power Electron.* **2017**, *32*, 6105–6118. [CrossRef]
5. Sun, W.; Xing, Y.; Wu, H.; Ding, J. Modified High-Efficiency LLC Converters with Two Split Resonant Branches for Wide Input-Voltage Range Applications. *IEEE Trans. Power Electron.* **2018**, *33*, 7867–7879. [CrossRef]
6. Qian, T.; Qian, C. An Adaptive Frequency Optimization Scheme for LLC Converter with Adjustable Energy Transferring Time. *IEEE Trans. Power Electron.* **2019**, *34*, 2018–2024. [CrossRef]
7. Yamamoto, T.; Konno, Y.; Sugimura, K.; Sato, T.; Bu, Y.; Mizuno, T. Loss Reduction of LLC Resonant Converter using Magnetocoated Wire. *IEEJ J. Ind. Appl.* **2019**, *8*, 51–56. [CrossRef]
8. Yamamoto, T.; Bu, Y.; Mizuno, T.; Yamaguchi, Y.; Kano, T. Loss Reduction of Transformer for LLC Resonant Converter Using a Magnetoplated Wire. *IEEJ J. Ind. Appl.* **2018**, *7*, 43–48. [CrossRef]
9. Kimura, S.; Nanamori, K.; Kawakami, T.; Imaoka, J.; Yamamoto, M. Allowable Power Analysis and Comparison of High Power Density DC-DC Converters with Integrated Magnetic Components. *IEEJ J. Ind. Appl.* **2017**, *6*, 463–472. [CrossRef]
10. Liu, B.; Liu, Z.; Liu, J.; Wu, T.; Wang, S. A feedforward control based power decoupling scheme for voltage-controlled grid-tied inverters. In Proceedings of the IEEE Applied Power Electronics Conference and Exposition, Long Beach, CA, USA, 20–24 March 2016; pp. 3328–3332.
11. Liu, X.; Dong, D.; Harfman-Todorovic, M.; Garaces, L. A PV Micro-inverter With PV Current Decoupling Strategy. In Proceedings of the IEEE Applied Power Electronics Conference and Exposition, Long Beach, CA, USA, 20–24 March 2016; pp. 3403–3408.
12. Roßkopf, A.; Bär, E.; Joffe, C. Influence of Inner Skin-and Proximity Effects on Conduction in Litz Wires. *IEEE Trans. Power Electron.* **2014**, *29*, 5454–5461. [CrossRef]
13. Ohnuma, Y.; Orikawa, K.; Itoh, J.I. A Single-Phase Current Source PV Inverter with Power Decoupling Capability Using an Active Buffer. In Proceedings of the IEEE Energy Conversion Congress and Exposition, Denver, CO, USA, 15–19 September 2013; pp. 3094–3101.

14. Watanabe, H.; Itoh, J.I. Zero Voltage Switching Scheme for Flyback Converter to Ensure Compatibility with Active Power Decoupling Capability. In Proceedings of the International Power Electronics Conference, Niigata, Japan, 20–24 May 2018; pp. 896–903.
15. Hurley, W.G.; Wolfle, W.H.; Breslin, J.G. Optimized Transformer Design: Inclusive of High-Frequency Effects. *IEEE Trans. Power Electron.* **1998**, *13*, 651–659. [CrossRef]

Article

Flux-Balance Control for LLC Resonant Converters with Center-Tapped Transformers

Yuan-Chih Lin, Ding-Tang Chen and Ching-Jan Chen *

Department of Electrical Engineering, National Taiwan University, Taipei, Taiwan; No. 1, Sec. 4, Roosevelt Rd., Taipei 10617, Taiwan
* Correspondence: chenjim@ntu.edu.tw; Tel.: +886-2-3366-3366 (ext. 348)

Received: 9 July 2019; Accepted: 19 August 2019; Published: 21 August 2019

Abstract: LLC resonant converters with center-tapped transformers are widely used. However, these converters suffer from a flux walking issue, which causes a larger output ripple and possible transformer saturation. In this paper, a flux-balance control strategy is proposed for resolving the flux walking issue. First, the DC magnetizing current generated due to the mismatched secondary-side leakage inductances, and its effects on the voltage gain are analyzed. From the analysis, the flux-balance control strategy, which is based on the original output-voltage control loop, is proposed. Since the DC magnetizing current is not easily measured, a current sensing strategy with a current estimator is proposed, which only requires one current sensor and is easy to estimate the DC magnetizing current. Finally, a simulation scheme and a hardware prototype with rated output power 200 W, input voltage 380 V, and output voltage 20 V is constructed for verification. The simulation and experimental results show that the proposed control strategy effectively reduces the DC magnetizing current and output voltage ripple at mismatched condition.

Keywords: LLC resonant converter; center-tapped transformer; flux walking; flux-balance control loop; magnetizing current estimation

---

1. Introduction

The LLC resonant converter is widely used in many different applications such as onboard chargers, server power systems, laptops, desktops, photovoltaic regeneration systems. Owing to the characteristics of zero-voltage-switching (ZVS) at the primary side and zero-current-switching (ZCS) at the secondary side, high efficiency and high power density of the LLC converter are achieved [1–7]. The half-bridge (HB) and full-bridge (FB) with the center-tapped transformer rectifier topologies shown in Figure 1 are the most commonly used topologies of the LLC resonant converter. Thanks to the center-tapped transformer, only two rectifying diodes are necessary at the secondary side [8–11]. Without the center-tap, it would be required to implement a full-wave rectifier.

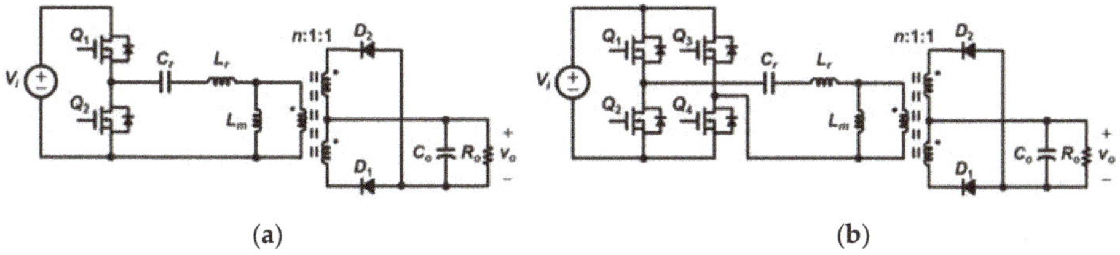

(a)   (b)

Figure 1. Different topologies of the LLC resonant converter (**a**) half-bridge and (**b**) full-bridge.

Figure 2a shows the key waveforms of the LLC resonant converter operating below the resonant frequency with the matching leakage inductances in both secondary side windings, where v_{gs1} and

v_{gs2} are the driving signals of the MOSFET Q_1 and Q_2, respectively. Figure 2a reveals that the current waveforms of the diode 1 (i_{D1}) and diode 2 (i_{D2}) are symmetrical, and energy flows through the diode 1 and diode 2 in the positive and negative cycles, respectively. In Figure 2a, the magnetizing current will have no DC component.

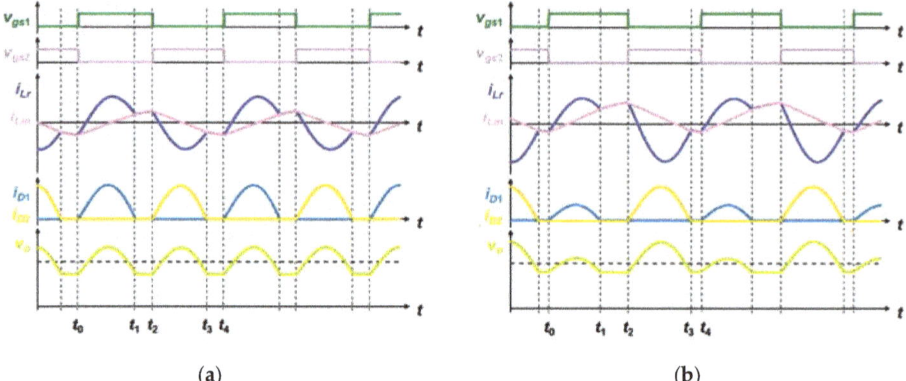

(a)   (b)

Figure 2. Key waveforms of the LLC resonant converter operating below the resonant frequency with the different leakage inductances in the secondary side windings (**a**) matched leakage inductances and (**b**) mismatched leakage inductances.

However, practical LLC resonant converters with center-tapped transformers contain a flux walking issue, which causes a larger output ripple and possible transformer saturation [12–14]. This is because the secondary side windings are usually not symmetrical in practical manufacture. The mismatched leakage inductances are generated in the secondary side windings, which causes an imbalance in the secondary side currents, resulting in a larger output voltage ripple. The secondary side currents imbalance will reflect upon the magnetizing inductance on the primary side. Therefore, the magnetizing current will have a DC component, which will lead to the flux walking and possible transformer core saturation.

Figure 2b shows the key waveforms of the LLC resonant converter operating below the resonant conditions with the mismatched leakage inductances in the secondary side windings. In Figure 2b, the current waveforms of the diode 1 and diode 2 are not symmetrical. The magnetizing current (i_{Lm}) will, therefore, contain a DC component. Since the resonant capacitance is in series with the primary side, the resonant current (i_{Lr}) will not have a DC component from charge balance concept. Besides, the mismatched condition considers not only the leakage inductances of the secondary side windings of the transformer but also the parasitic inductances of the printed-circuit-board (PCB) traces on the secondary side.

To improve the flux walking issue in the LLC resonant converter, improved winding structures for the secondary side were proposed [12–14]. The coupling coefficients between the primary side and two secondary sides were increased to mitigate the flux walking issue [12]. The further improved method is used in the Bifilar winding structure to overcome this problem [13]. The flux distribution of the non-symmetrical structure of the secondary side windings were analyzed in [14]. However, they cannot consider the mismatch problems caused by the PCB circuit traces.

According to abovementioned issue, a flux-balance control strategy, which is based on the original output-voltage control loop, is proposed in this paper. The flux-balance control is added to improve the magnetizing current imbalance problem caused by the secondary side mismatches.

Besides, the DC magnetizing current of the LLC resonant converter is difficult to sense directly. An indirect method to sense the magnetizing current was used [15], which involved sensing the currents at the primary and secondary side, simultaneously and subtracting them to obtain the magnetizing current. Then, the DC component was obtained using a low-pass filter. This solution, however, required several current sensing devices, which increased the circuit cost; moreover, the low-pass filter produced a slow dynamic response. Therefore, this paper proposes a simple magnetizing current sensing strategy with a DC current estimation scheme to overcome the abovementioned issue.

Moreover, for small-signal model, the mathematical methods for deriving the small-signal dynamic model of the LLC resonant converter were developed [16–19]. However, these works focused only on the switching frequency to output voltage transfer function. Besides, matching with circuit ac sweep simulation occurs only under specific operating conditions. Nevertheless, system identification [20,21] is a useful method to obtain the small-signal model of the system without using any complex mathematical model. Therefore, the system identification [21] is used to obtain the small-signal models of the LLC resonant converter for controllers design in this paper.

The remainder of this paper is organized as follows. Section 2 analyzes the mismatched leakage inductances effects of the secondary side on the DC magnetizing current and voltage gain. Section 3 describes the proposed control strategy with the magnetizing current sensing and the DC magnetizing current estimation. Section 4 describes the controller designs for the flux-balance loop and output-voltage loop, which are based on the transfer functions obtained by using the system identification tool of MATLAB (R2018b, MathWorks, Natick, MA, USA). Section 5 presents the simulated and experimental results, to verify the effectiveness of the proposed control strategy. Finally, Section 6 provides the conclusions.

2. Analysis of Secondary Side Mismatched Leakage Inductances Effects

2.1. Analysis of the Magnetizing Current DC Value

Figure 3 shows the DC current path in the LLC resonant tank when the leakage inductances at the secondary side are mismatched. It is assumed that the leakage inductance of the positive-cycle loop at the secondary side ($L_{lk2,pos}$) is smaller than the leakage inductance of the negative-cycle loop at the secondary side ($L_{lk2,neg}$). According to Kirchhoff's current law (KCL), the relationship among i_{Lr}, i_{Lm}, and i_{Dx} can be expressed as follows

$$i_{Lr} = i_{Lm} + \frac{i_{Dx}}{n} \tag{1}$$

$$x = \begin{cases} 1 : during the positive cycle \\ 2 : during the negative cycle \end{cases} \tag{2}$$

where n is the turns ratio of the transformer. Since the resonant capacitance is in series with the input of the transformer, the DC current of the resonant inductor is zero. That is, the average resonant inductor current in a switching period is zero in steady-state, and can be expressed as follows

$$I_{Lr,DC} = \frac{1}{T_s} \int_0^{T_s} i_{Lr} dt = \frac{1}{T_s} \int_0^{T_s} \left(i_{Lm} + \frac{i_{Dx}}{n} \right) dt = 0 \text{A}. \tag{3}$$

Based on (3), the relationship between the magnetizing DC current and difference in the DC currents of both diodes can be expressed as follows

$$\begin{aligned} I_{Lm,DC} &= \frac{1}{T_s} \int_0^{T_s} i_{Lm} dt = -\frac{1}{nT_s} \int_0^{T_s} i_{Dx} dt \\ &= -\frac{1}{nT_s} \left(\int_0^{T_s/2} i_{D1} dt - \int_{T_s/2}^{T_s} i_{D2} dt \right) = \frac{1}{n} (I_{D2,DC} - I_{D1,DC}) \end{aligned} \tag{4}$$

where $I_{D1,DC}$ and $I_{D2,DC}$ are the DC currents of the diodes at secondary side. Equation (4) shows that the magnetizing DC current is proportional to the difference between $I_{D1,DC}$ and $I_{D2,DC}$ with the turns ratio n. Therefore, the largest $I_{Lm,DC}$ would be induced for the largest current mismatch in the secondary side condition.

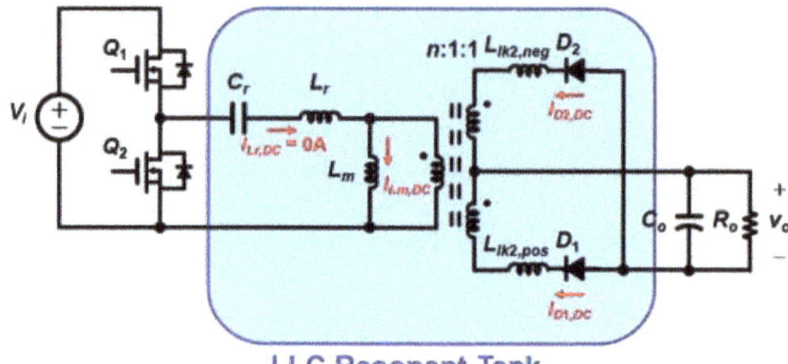

Figure 3. The DC current path in the LLC resonant tank when the leakage inductances at the secondary side are mismatched ($L_{lk2,pos} < L_{lk2,neg}$).

2.2. Analysis of the Voltage Gain under Mismatched Condition

A non-ideal transformer equivalent circuit can be expressed by the model shown in Figure 4a, which is called the "T model" [22,23]. In Figure 4a, the leakage inductances are distributed at the primary and secondary sides, separately. This circuit is not suitable for the analysis of the LLC resonant converter. On the other hand, Figure 4b shows the "L model" [22,23]. In this model, the leakage inductance at the secondary side is removed. Therefore, it is suitably used in the LLC resonant tank for analysis.

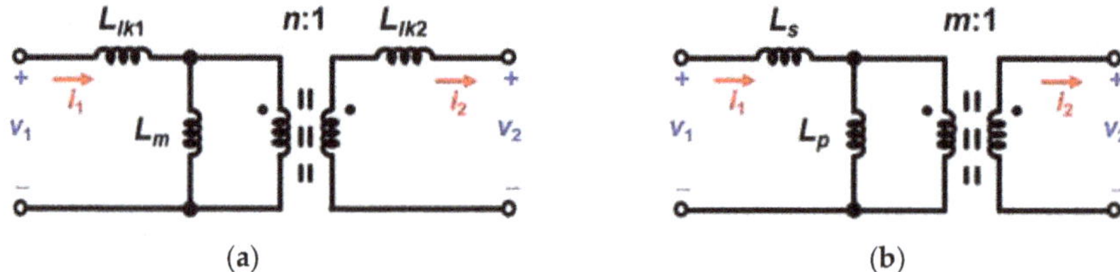

Figure 4. Different equivalent circuit models of the transformer (**a**) T model and (**b**) L model.

For center-tapped applications, the L model of the transformer can be separated into the positive and negative cycle models. For a matched condition, the parameters of the L model for the positive and negative cycle will be the same. This will not be the case for the mismatched condition because the leakage inductances at the secondary side would affect the resonant parameters during the positive and negative cycles.

Figure 5 shows the equivalent circuit models of the LLC resonant tank for the positive (lower left) and negative (lower right) cycles. The relative parameters of the L model during the positive and negative cycles can be derived from Figure 4 and is presented as follows

$$m_h = \frac{nL_m}{L_m + n^2 L_{lk2,h}} \tag{5}$$

$$L_{p,h} = \frac{L_m{}^2}{L_m + n^2 L_{lk2,h}} \tag{6}$$

$$L_{r,h} = L_{ext} + L_{lk1} + L_m \| n^2 L_{lk2,h} \tag{7}$$

$$h = \begin{cases} pos : during\,the\,positive\,cycle \\ neg : during\,the\,negative\,cycle \end{cases} \tag{8}$$

where L_m indicates the magnetizing inductance of the transformer, L_{lk1} and $L_{lk2,h}$ express the leakage inductances at the primary and secondary sides of the transformer, respectively, m_h indicates the equivalent turns ratio of the L model, L_{ext} represents the external resonant inductance, $L_{p,h}$ expresses the equivalent paralleled inductance of the L model, and $L_{r,h}$ is the total equivalent resonant inductance of the L model. Therefore, the parameters of the L model during the positive and negative cycles can be obtained using (5)–(8).

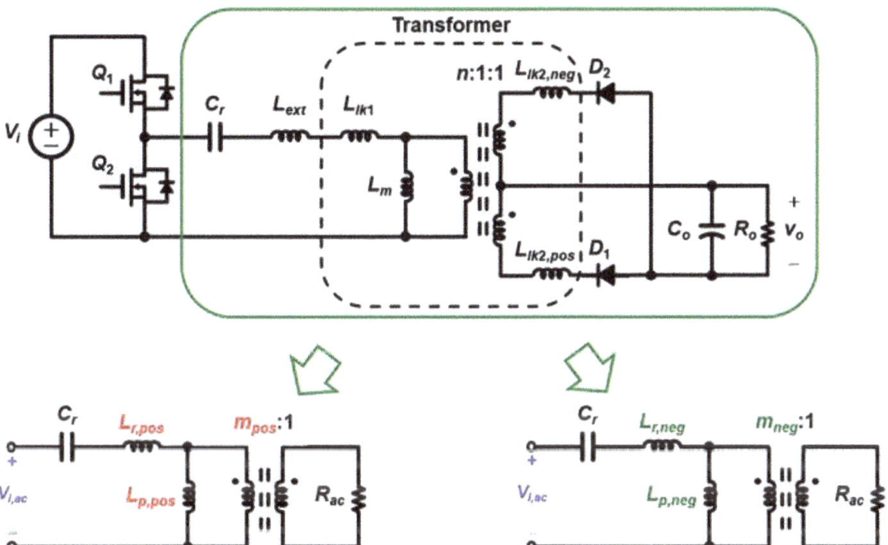

Figure 5. Equivalent circuit models of the resonant tank during the positive cycle (lower left) and negative cycle (lower right).

The voltage gain during the positive and negative cycles can be expressed as follows

$$M(f_{n,h}) = \frac{m_h V_o}{V_i} = \frac{1}{\sqrt{\left[1 + \frac{1}{L_{n,h}}\left(1 - \frac{1}{f_{n,h}^2}\right)\right]^2 + \left[Q_h\left(f_{n,h} - \frac{1}{f_{n,h}}\right)\right]^2}} \tag{9}$$

$$f_{r,h} = \frac{1}{2\pi \sqrt{L_{r,h}C_r}} \tag{10}$$

$$Q_h = \frac{1}{m_h^2 R_{ac}}\sqrt{\frac{L_{r,h}}{C_r}} \tag{11}$$

$$R_{ac} = \frac{8}{\pi^2}R_o \tag{12}$$

$$f_{n,h} = \frac{f_s}{f_{r,h}} \tag{13}$$

$$L_{r,h} = \frac{L_{p,h}}{L_{r,h}} \tag{14}$$

where f_r indicates the resonant frequency, C_r is the resonant capacitance, Q_h is the qualify factor, R_{ac} is the equivalent ac load resistance, R_o is the load resistance, f_s is the switching frequency, $f_{n,h}$ is the normalized frequency, and $L_{n,h}$ is the ratio between $L_{p,h}$ and $L_{r,h}$.

Based on (9)–(14), the voltage gain curves during the positive and negative cycles can be drawn as shown in Figure 6. The solid lines represent the nominal Q-value condition and dashed lines express the high Q-value condition. Assuming the normal leakage inductance $L_{lk,2n} = L_{lk,2,pos} = L_{lk,2,neg}$ at the secondary side, i.e., at matched condition, $f_{n,pos}$ and $f_{n,neg}$ would be one as indicated by the curve **a** (i.e., maroon solid line) with the normal operating point indicated by the sky blue circle in Figure 6. When operating under the mismatched condition and assuming the leakage inductances at the secondary side are satisfying $L_{lk,2n} = L_{lk,2,pos} < L_{lk,2,neg}$, the voltage gain during the positive cycle follows the original curve **a** (i.e., maroon solid line); during the negative cycle, the voltage gain curve moves toward the left as shown by the curve **b** (i.e., light blue solid line). Under the mismatched condition, the voltage gain also operates at $M = 1.05$, because of the voltage loop regulation, and the operating point shifts to the point indicated by the pink circle.

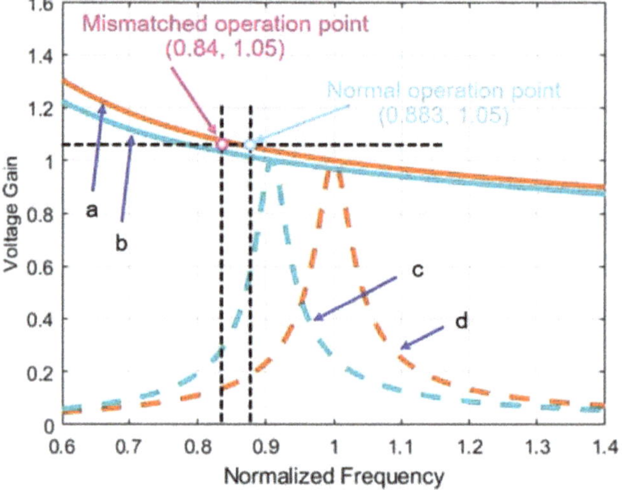

Figure 6. Voltage gain curves during the positive and negative cycles, curve **a**: Nominal Q-value condition with the matched leakage inductance $L_{lk,2n}$ at the secondary side; curve **b**: Nominal Q-value condition with the mismatched leakage inductance $L_{lk,2,neg}$ at the secondary side during the negative cycle; curve **c**: High Q-value condition with the matched leakage inductance $L_{lk,2n}$ at the secondary side; curve **d**: High Q-value condition with the mismatched leakage inductance $L_{lk,2,neg}$ at the secondary side during the negative cycle.

The operations of the curve **a** (i.e., maroon solid line) during the positive cycle and curve **b** (i.e., light blue solid line) during the negative cycle can be mapped to time-domain operated waveforms shown in Figure 2b. They are matched precisely between the voltage gains and time domain waveforms.

3. Proposed Flux-Balance Control Architecture

3.1. Description of the Proposed Control Architecture

Figure 7 shows the block diagram of the proposed control architecture to solve the flux walking issue caused by the mismatched condition on the secondary side. The proposed control is based on the original output-voltage control loop with the addition of the flux-balance loop. The flux-balance loop includes the sampling setup for the magnetizing current, a DC current estimator, a flux-balance loop controller, and a variable-frequency-variable-duty-pulse-width-modulator (VFVDPWM). The output of the voltage loop controller controls the switching period of the MOSFETs Q_1 and Q_2. The output of the flux-balance loop controller offsets the duty ratio of the MOSFETs Q_1 and Q_2 from 0.5. The input of the flux-balance loop controller is the error between $i_{Lm,DC}$ and $i_{Lm,DC,ref}$, which is zero in steady-state. Thus, the flux-balance loop forces the DC magnetizing current to zero and solves the flux walking issue. The DC estimator estimates the DC magnetizing current, and the i_{Lm} sampling scheme samples relative information of the magnetizing current from the resonant current i_{Lr}. The relationship between the DC magnetizing current and the duty ratio are not direct, because the resonant capacitance is in series between the half-bridge switches and the input of the transformer. According to the proposed flux-balance control, when the duty is regulated, the resonant capacitance to hold the charge balance in steady-state, the DC current of the resonant inductance, therefore, keeps zero, and the rectifier diodes turned off time, therefore, be changed when the duty ratio be regulated. Finally, the magnetizing DC current would be regulated. Detailed descriptions of the functional blocks of the magnetizing current sampling scheme, the DC current estimator, and the VFVDPWM are provided below.

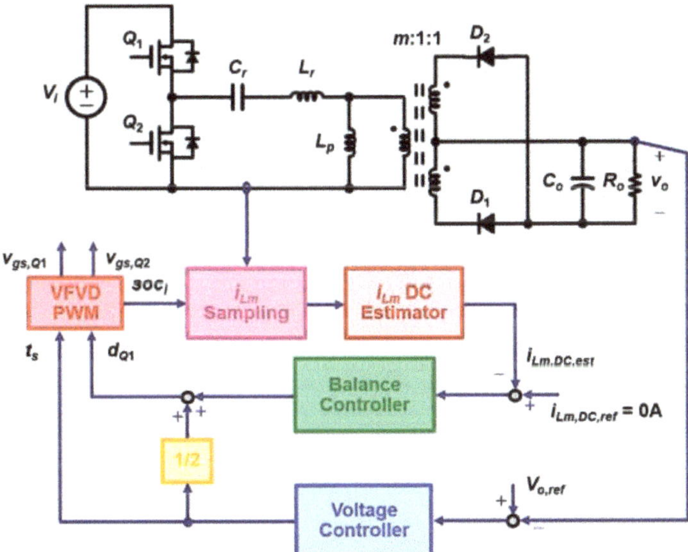

Figure 7. Overall block diagram of the proposed control architecture.

3.2. Magnetizing Current Sampling Scheme

Figure 8 shows the magnetizing current sampling scheme, where i_{Lr} is the resonant current, i_{Lm} is the magnetizing current, $i_{Lm,p}$, and $i_{Lm,n}$ are the peak values of the magnetizing current during the positive and negative cycles, respectively. The magnetizing current of the LLC resonant converter cannot be measured directly because the magnetizing inductance is an equivalent element in the transformer. However, owing to the LLC resonant converter usually operates with the switching frequency below the resonant frequency, the resonant current is equal to the magnetizing current when the diodes on the secondary side are turned off. As shown in Figure 8, during these intervals, the peak values of the magnetizing current during the positive and negative cycles occur when the respective MOSFET Q_1 and Q_2, are turned off. According to the previous statement, the sampling pluses of $SOC_{i,p}$ and $SOC_{i,n}$ can be applied from the VFVDPWM, which are generated when the driving signals of MOSFET Q_1 and Q_2 are turned off, respectively. After that, $i_{Lm,p}$ and $i_{Lm,n}$ can be sampled during each switching period.

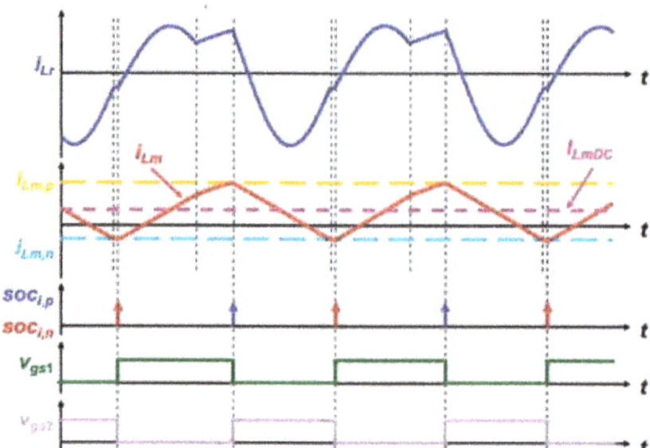

Figure 8. Magnetizing current sampling timing, where $i_{Lm,p}$ ($i_{Lm,n}$) is sampled at the timing of v_{gs1} (v_{gs2}) from high to low.

3.3. DC Value Estimation of the Magnetizing Current

This paper proposes the DC value estimation scheme to obtain the DC magnetizing current, which will be used as one of the inputs of the flux-balance controller. Figure 9 shows the magnetizing current waveform (red line), which is similar to a triangular wave (shown by the blue dashed line). The magnetizing DC current can be estimated after the $i_{Lm,p}$ and $i_{Lm,n}$ are sampled, and is expressed as follows

$$
\begin{aligned}
i_{Lm,DC,est}[n] &= \mathrm{DC\,value\,of\,triangular\,wave} \\
&= \tfrac{1}{2}\left(i_{Lm,p}[n] + i_{Lm,n}[n]\right) \\
&\approx \frac{1}{T_s}\int_{0}^{T_s} i_{Lm}\,dt
\end{aligned}
\tag{15}
$$

where $i_{Lm,DC,est}[n]$ represents the estimated magnetizing DC current. Equation (15) reveals that the estimated magnetizing DC current can be approximated as the sum of $i_{Lm,p}[n]$ and $i_{Lm,n}[n]$, divided by 2, which is a very simple method to obtain the magnetizing DC current. The block diagram of the DC magnetizing current estimator is shown in Figure 10.

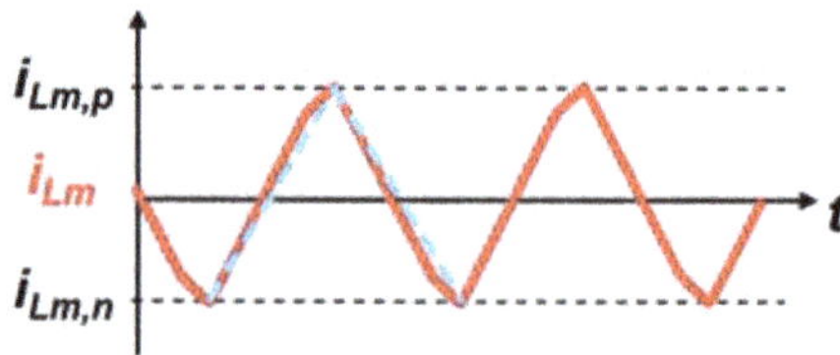

Figure 9. The magnetizing current waveform (red line) and the approximate triangular waveform (blue dashed line).

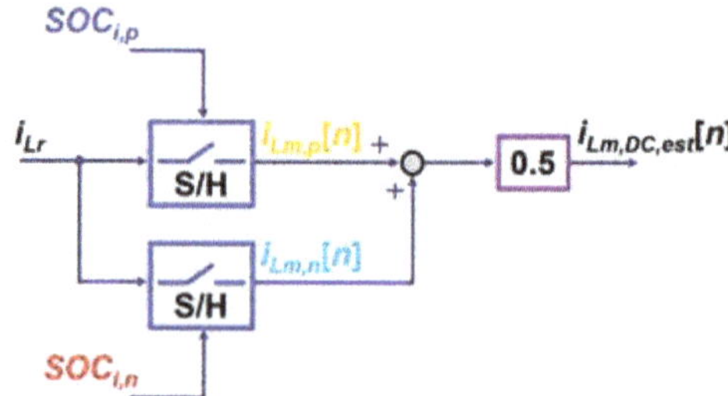

Figure 10. Block diagram of the magnetizing DC current estimator.

3.4. Variable-Frequency-Variable-Duty-Pulse-Width-Modulator

In conventional LLC resonant converters, the output voltage regulation is achieved by controlling the switching frequency and maintaining the duty ratios of MOSFET Q_1 and Q_2 at 50%. The proposed flux-balance control scheme utilizes the controlled duty ratio for MOSFET Q_1 for the magnetizing DC current regulation. Figure 11 shows the operating timing of the VFVDPWM, where v_{carr} is the carrier waveform, d_{Q1} is the duty ratio control signal of MOSFET Q_1, t_s is the switching period control signal. In Figure 11, the slope of v_{carr} is fixed. t_s can therefore control the magnitude of v_{carr} to control the switching period. d_{Q1}, which is equal to $0.5t_s$ in the matched condition, controls the duty ratio of MOSFET Q_1 for the flux-balance loop regulation. $SOC_{i,p}$ is generated when d_{Q1} equals v_{carr} and $SOC_{i,n}$ is generated when t_s equals v_{carr}. Thus, the control signal and magnetizing current sampling pulses of the proposed flux-balance loop can be obtained from Figure 11.

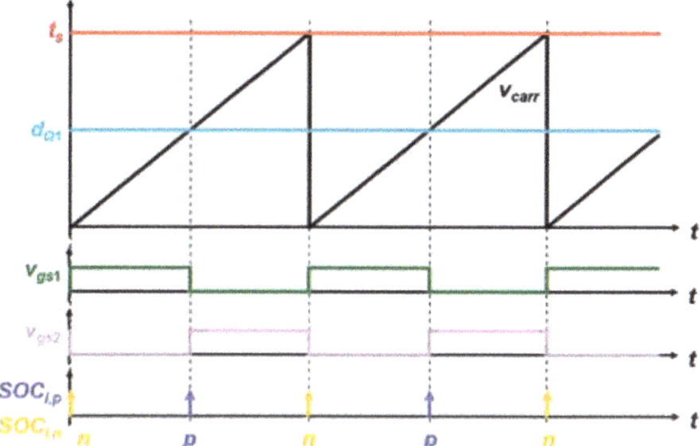

Figure 11. VFVDPWM operating timing.

4. Small-Signal Model and Controllers Design

4.1. Small-Signal Models Built Using System Identification

Unlike the pulse-width-modulation (PWM) converter, the LLC resonant converter is a complex nonlinear system. Therefore, it is difficult to obtain its small-signal model through mathematical derivation. Previously, researchers had developed the mathematical methods for deriving the small-signal dynamic model of the LLC resonant converter [16–19]. However, these works focused only on the matching with circuit ac sweep simulation occurs only under specific operating conditions. At the same time, the flux-balance loop proposed in this paper, there is no more literature to discuss. The system identification tool of MATLAB [21] is a useful tool to obtain the small-signal model transfer function without using any complex mathematical model derivation. Figure 12 shows the processing flow windows of the system identification of MATLAB. The transfer function of the system can be obtained from the simulated or measured system data such as time response or frequency response. Therefore, the system identification tool of MATLAB [21] is used to obtain the small-signal model for controller design, in this paper.

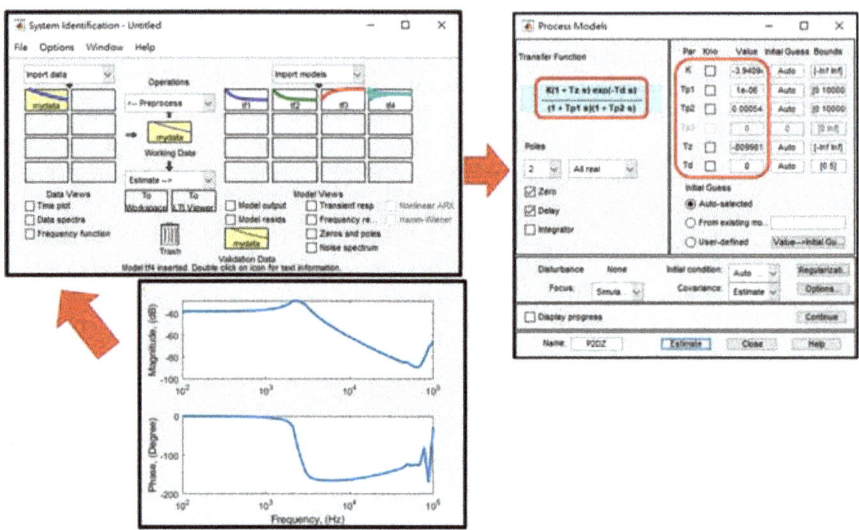

Figure 12. Processing flow windows of the system identification tool of MATLAB.

4.2. Flux-Balance Loop Controller Design

Figure 13 shows the control block diagram of the flux-balance loop, for which the loop gain can be expressed as

$$T_b(s) = G_{comp,b}(s)G_{PWM,b}(s)G_{bc}(s) \tag{16}$$

where $G_{bc}(s)$ represents the transfer function of the controlled plant, which is the duty ratio of MOSFET Q_1 (d_1-tilde) to the magnetizing DC current ($i_{Lm,DC}$-tilde), $G_{PWM,d}(s)$ represents the transfer function of the duty ratio control signal ($v_{con,d}$-tilde) to the duty ratio of MOSFET Q_1, and $G_{comp,b}(s)$ indicates the controller of the flux-balance loop.

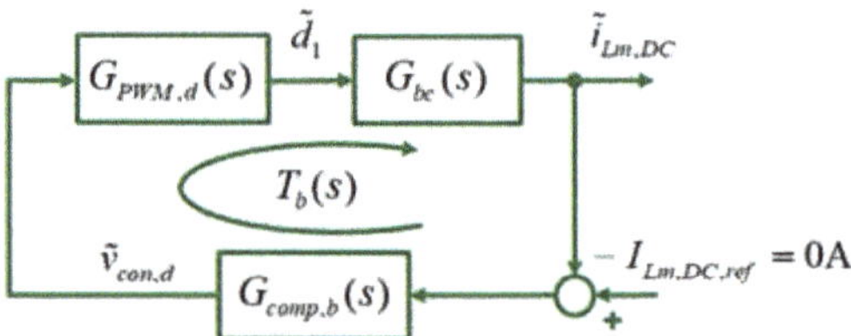

Figure 13. Control block diagram of the flux-balance loop.

Using the system identification tool of MATLAB, the uncompensated loop gain from the duty ratio control signal to the magnetizing DC current under full load operation condition can be expressed as follows. The parameters of the LLC resonant converter are shown in Table 1, which will be shown in Section 5.

$$\frac{\tilde{i}_{Lm,DC}}{\tilde{v}_{con,d}} = G_{PWM,d}(s)G_{bc}(s) \\ = \frac{1.452\times10^9}{s^2+487.1\times10^3s+41.49\times10^9}. \tag{17}$$

The characteristic equation in (17) has two real poles at 110 krad/s (17.5 kHz) and 377 krad/s (60 kHz), respectively. Figure 14a shows the bode plots of the uncompensated flux-balance loop gains obtained from the mathematical model in (17) (blue dashed line) and Simplis circuit simulation (red solid line). Figure 14a reveals that the dominate pole is at 17.5 kHz. The controller of the flux-balance loop $G_{comp,b}(s)$ can, therefore, be chosen as a PI-type controller and can be expressed as follows

$$G_{comp,b}(s) = K_{pb}\frac{s+z_b}{s}$$
$$= 20 \cdot \frac{s+62.8\times10^3}{s} \tag{18}$$

where K_{pb} is the DC gain, which is determined according to the crossover frequency sets as $f_{c,b} = 10$ kHz, and z_b is the zero, which is set at 17.5 kHz for pole/zero cancellation. The phase margin (PM) is set as 72° to ensure stability. The bode plot of the compensated flux-balance loop gain, after the addition of the controller, is shown in Figure 14b.

Table 1. Parameters of the LLC resonant converter.

Symbol	Description	Quantity
V_i	input voltage	380 V
V_o	output voltage	20 V
$P_{o,rated}$	rated output power	200 W
n	transformer turns ratio	10
L_m	magnetizing inductance	310 µH
L_{lk1}	leakage inductance in the primary side	6.386 µH
$L_{lk2,pos}$	leakage inductance in the secondary side during the positive cycle	53 nH
$L_{lk2,neg}$	leakage inductance in the secondary side during negative cycle (matched condition)	53 nH
$L_{lk2,neg}$	leakage inductance in the secondary side during negative cycle (mismatched condition)	167.77 nH
L_{ext}	external resonant inductance	42 µH
C_r	resonant capacitance	20 nF
C_o	output capacitance	1000 µF
r_{Co}	equivalent-series-resistance of the output capacitance	40 mΩ

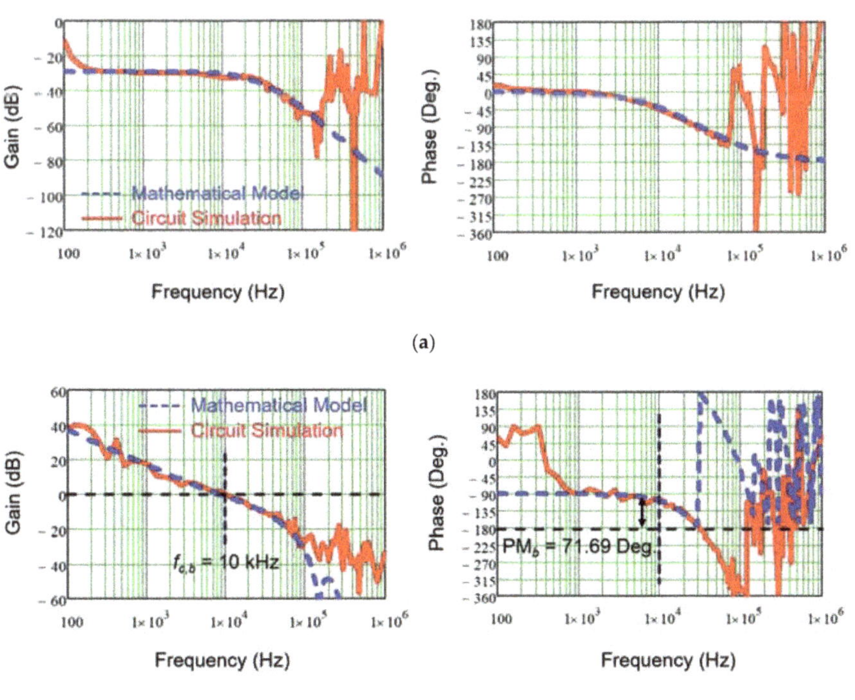

Figure 14. Bode plot of the flux-balance loop gain (**a**) uncompensated and (**b**) compensated.

4.3. Output-Voltage Loop Controller Design

Figure 15 shows the control block diagram of the output-voltage loop, the loop gain of which can be expressed as follows

$$T_v(s) = G_{comp,v}(s)G_{PWM,f}(s)G_{fv}(s) \tag{19}$$

where $G_{fv}(s)$ represents the transfer function of the controlled plant, which is the switching frequency of MOSFET Q_1 (f-tilde) to the output voltage (v_o-tilde), $G_{PWM,d}(s)$ represents the transfer function of the switching frequency control signal ($v_{con,f}$-tilde) to the switching frequency of MOSFET Q_1, and $G_{comp,v}(s)$ is the controller of the output-voltage loop.

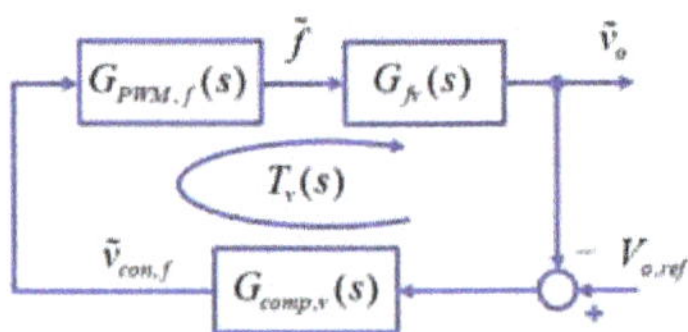

Figure 15. Control block diagram of the output voltage loop.

Using the system identification tool of MATLAB, the uncompensated loop gain from the switching frequency control signal to the output voltage under full load operation condition can be expressed as follows. The parameters of the LLC resonant converter are shown in Table 1.

$$\begin{aligned}\frac{\tilde{v}_o}{\tilde{v}_{con,f}} &= G_{PWM,f}(s)G_{fv}(s)\\ &= \frac{594\times(s+30.97\times10^3)}{s^2+42.84\times10^3s+795.4\times10^6}.\end{aligned} \tag{20}$$

The characteristic equation in (20) has a complex pole pair at $21.42 \pm j18.346$ krad/s ($3.41 \pm j2.9$ kHz) and a zero at 30.97 krad/s (4.93 kHz). Figure 16a shows the bode plots of the uncompensated output-voltage loop gains obtained from the mathematical model in (20) (blue dashed line) and Simplis circuit simulation (red solid line). Figure 16a reveals that the phase down to $-90°$, due to the zero, is very close to complex pole pair. Thus, (20) can be simplified as a first-order system. The controller of the output voltage $G_{comp,v}(s)$ can also be chosen as a PI-type controller and can be expressed as follows

$$\begin{aligned}G_{comp,v}(s) &= K_{pv}\frac{s+z_v}{s}\\ &= 70\cdot\frac{s+18.85\times10^3}{s}\end{aligned} \tag{21}$$

where K_{pv} is the DC gain, which is determined according to the crossover frequency sets as $f_{c,v} = 6$ kHz, z_v is the zero, which is set at 18.85 kHz for pole/zero cancellation, and the phase margin at $f_{c,v}$ is 78.33° to ensure stability. The bode plot of the compensated output-voltage loop gain, after the addition of the controller, is shown in Figure 16b.

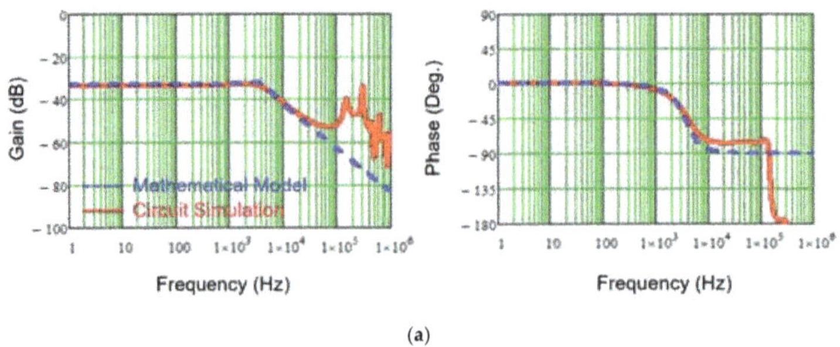

Figure 16. *Cont.*

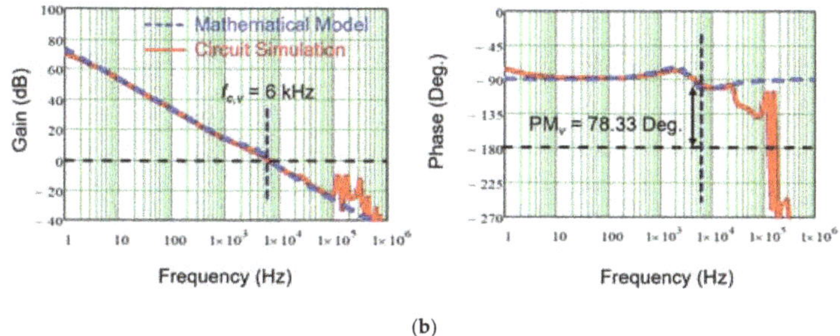

(b)

Figure 16. Bode plot of the output-voltage loop gain (**a**) uncompensated and (**b**) compensated.

5. Simulation and Experimental Verification

To verify the proposed approach, Simplis software was used to construct the simulation, and an experimental platform was built as shown in Figure 17. In this paper, the center-tapped model constructed in Simplis is similar to Figure 5. Two ideal transformers, where the primary side is connecting in parallel, and the secondary side connecting in series, are used. The magnetizing inductance and leakage inductances are added in the primary side and secondary sides, respectively. The advantage of this method is that all of branches current and nodes voltages can easily be measured for observation and analysis.

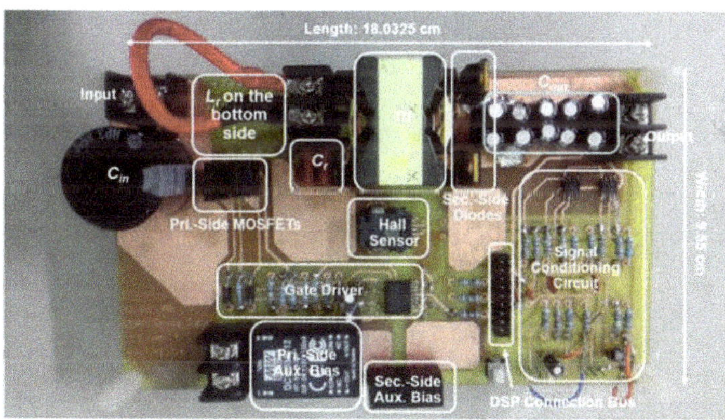

Figure 17. Experimental prototype of LLC resonant converter.

The proposed flux-balance control strategy was implemented by using Texas Instrument C2000 Piccolo 28035 digital signal processor. For experimental equipment, Agilent Technologies InfiniiVision DSO-X 3054A oscilloscope (BW = 500 MHz) was used; Keysight Technologies 1147B current probe (BW = 50 MHz) was used for i_{L_r} measurement, and Sapphire Instrument LDP-6002 (BW = 25 MHz) differential voltage probes were used for differential voltage measurement (v_{gs1}). Table 1 lists the related parameters of the LLC resonant converter. The leakage inductance at the secondary side, during the negative cycle, covers the matched condition ($L_{lk2,neg}$ = 53 nH) and mismatched condition ($L_{lk2,neg}$ = 167.77 nH). The set switching frequency is less than the resonant frequency, i.e., $f_s < f_r$, for an input voltage V_i = 380 V.

5.1. Steady-State Operation

Figure 18 shows the simulated and experimental results of the LLC resonant converter at full load in steady-state operation with the matched secondary-side leakage inductances, i.e., $L_{lk2,pos} = L_{lk2,neg}$

= 53 nH. In Figure 18a, the peak-to-peak ripple of the output voltage is approximately 734 mV. The conduction times of the secondary side diodes are 2.96 µs. The resonant frequency is f_r = 168.9 kHz, and switching frequency is f_s = 134.78 kHz. In Figure 18b, the peak-to-peak ripple of the output voltage is 700 mV and conduction times of the secondary-side diodes are approximately 3.14 µs during the positive and negative cycles. Therefore, the resonant frequency is f_r = 159.23 kHz and switching frequency is f_s = 130.11 kHz. The simulated and experimental results show that the magnetizing current is almost balanced between horizontal axis, i.e., $I_{Lm,DC}$ = 0 A, but some parasitic effects in the secondary side of the hardware cause a mismatch between the simulated and experimental results.

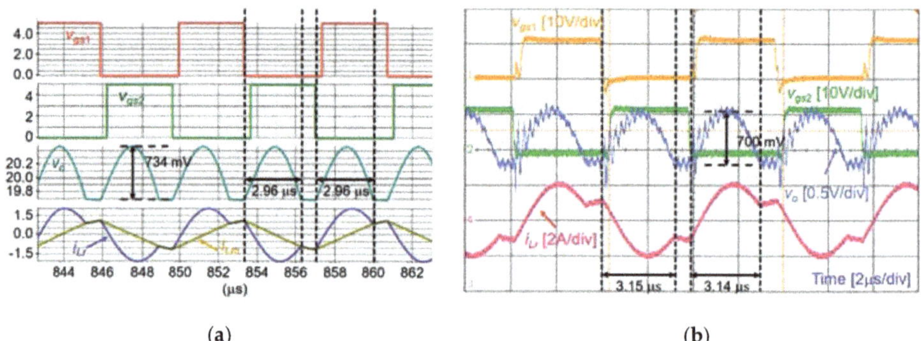

(a)   (b)

Figure 18. Matched secondary side leakage inductances at full load in steady-state operation (**a**) simulation results and (**b**) experiment results.

Figure 19 shows the simulated and experimental results of the LLC resonant converter at steady-state, when operating at full load with the mismatched secondary side leakage inductances, i.e., $L_{lk2,pos}$ = 53 nH and $L_{lk2,neg}$ = 167.77 nH. In Figure 19a, the peak-to-peak ripples of the output voltage during the positive and negative cycles are 744 mV and 881 mV, respectively. The conduction time of the secondary side diode during the positive cycle is 2.83 µs. Therefore, the resonant frequency during the positive cycle is $f_{r,pos}$ = 176.67 kHz. The conduction time of the secondary side diode during the negative cycle is 3.15 µs. Hence, the resonant frequency during the negative cycle is $f_{r,neg}$ = 158.73 kHz. The switching frequency is f_s = 127.98 kHz and magnetizing DC current is $I_{Lm,DC}$ = 436 mA. In Figure 19b, the peak-to-peak ripples of the output voltage during the positive and negative cycles are 750 mV and 1 V, respectively. The conduction time of the secondary side diode during the positive cycle is 3 µs. Hence, the resonant frequency during positive cycle is $f_{r,pos}$ = 166.67 kHz. The conduction time of the secondary side diode during the negative cycle is 3.2 µs; therefore, the resonant frequency during the negative cycle is $f_{r,neg}$ = 156.25 kHz. The switching frequency is f_s = 127.01 kHz. The magnetizing DC current can be estimated as $I_{Lm,DC,est}$ = 400 mA using (15), which is approximately the same as the simulated result, although the experimental result does not measure the magnetizing DC current directly.

Figure 20 shows the simulated and experimental results of the LLC resonant converter at full load in steady-state with the flux-balance loop control, for the same mismatched condition as that in Figure 19. In Figure 20a, the peak-to-peak ripples of the output voltage during the positive and negative cycles are 780 mV and 774 mV, respectively. The conduction time of the secondary side diode during the positive cycle is 2.97 µs; therefore, the resonant frequency during positive cycle is $f_{r,pos}$ = 168.35 kHz. The conduction time of the secondary side diode during the negative cycle is 3.04 µs; hence, the resonant frequency during the negative cycle is $f_{r,neg}$ = 164.47 kHz. The switching frequency is f_s = 125.37 kHz and magnetizing DC current is $I_{Lm,DC}$ = 19 mA. In Figure 20b, the peak-to-peak ripples of the output voltage during the positive and negative cycles are 750 mV and 740 mV, respectively. The conduction time of the secondary side diode during the positive cycle is 3.15 µs; hence, the resonant frequency during positive cycle is $f_{r,pos}$ = 158.73 kHz. The conduction time of the secondary side diode during the negative cycle is 3.2 µs; therefore, the resonant frequency during negative cycle is $f_{r,neg}$ = 156.25 kHz. The switching frequency is f_s = 126.21 kHz. The magnetizing DC current can be estimated to be $I_{Lm,DC,est}$ = 20 mA, using (15). The magnetizing DC current estimated from the experimental result is close to the value obtained from the simulations, which confirms the effectiveness of the approach proposed in this paper.

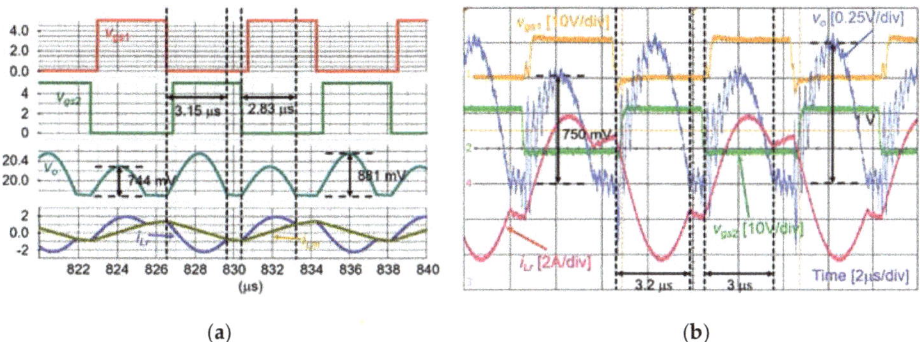

(a)   (b)

Figure 19. Mismatched secondary side leakage inductances at full load in steady-state operation (**a**) simulation results and (**b**) experiment results.

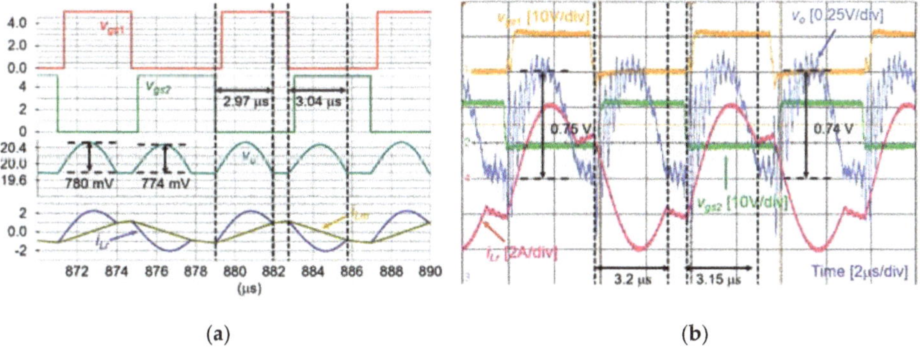

(a)   (b)

Figure 20. Mismatched secondary side leakage inductances at full load in steady-state operation with the flux-balance loop control (**a**) simulation results and (**b**) experiment results.

5.2. Dynamic-State Operation

Figure 21 shows the simulated and experimental results of the transient response of the LLC resonant converter, which compensates for the mismatched secondary side leakage inductances using the flux-balance loop, for load changes from 50% to 70%. The transient time is approximately 200 µs, as shown in the simulated and experimental results. Figure 21 reveals that the voltage loop and the flux-balance loop operate simultaneously, without affecting each other in the transient state. Figure 22 shows the simulated and experimental results of the transient response of the LLC resonant converter, which compensates for the mismatched secondary side leakage inductances with the flux-balance loop, for load changes from 70% to 50%. The simulated and experimental results show that the transient time is 200 µs. Figure 22 shows that the voltage loop and flux-balance loop do not affect each other, even during the load step down. Figure 23 shows the experimental results of the disabled and enabled flux-balance loop at 70% load, considering the mismatched secondary side leakage inductances. When the flux-balance loop is disabled, the maximum peak-to-peak ripple of the output voltage is

818.75 mV. Once the flux-balance loop is enabled, the maximum peak-to-peak ripple of the output voltage reduces to 731.25 mV.

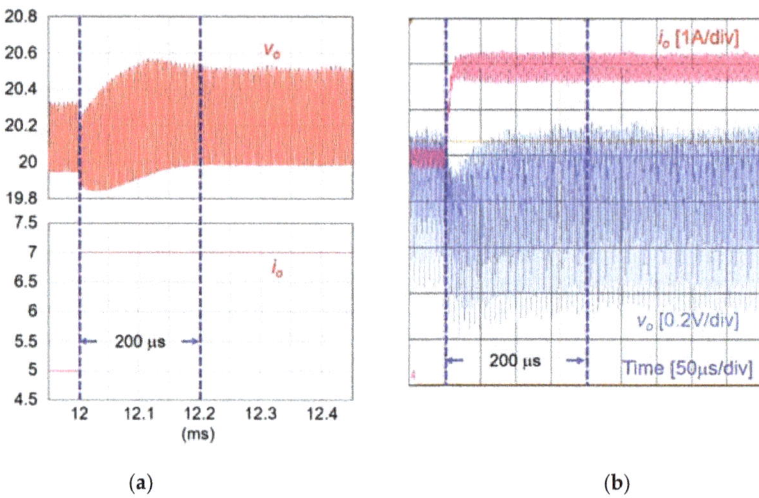

(a)

(b)

Figure 21. Transient response of load change from 50% to 70% with the mismatched secondary side leakage inductances and the flux-balance loop control (**a**) simulation results and (**b**) experiment results.

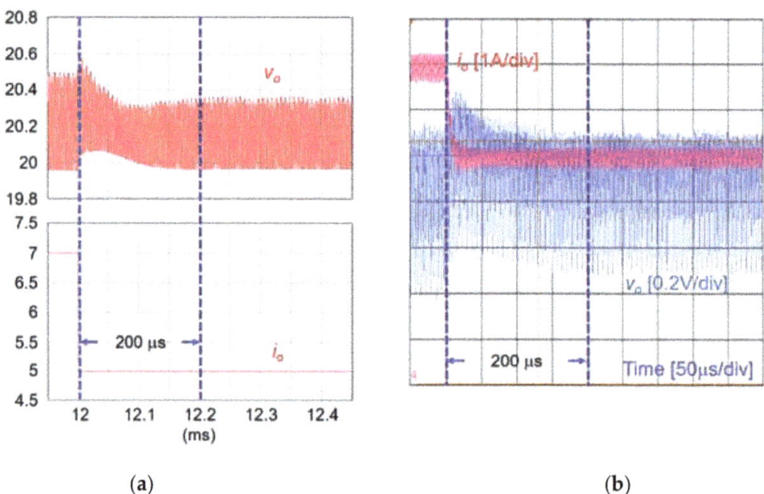

(a)

(b)

Figure 22. Transient response of load change from 70% to 50% with the mismatched secondary side leakage inductances and the flux-balance loop control (**a**) simulation results and (**b**) experiment results.

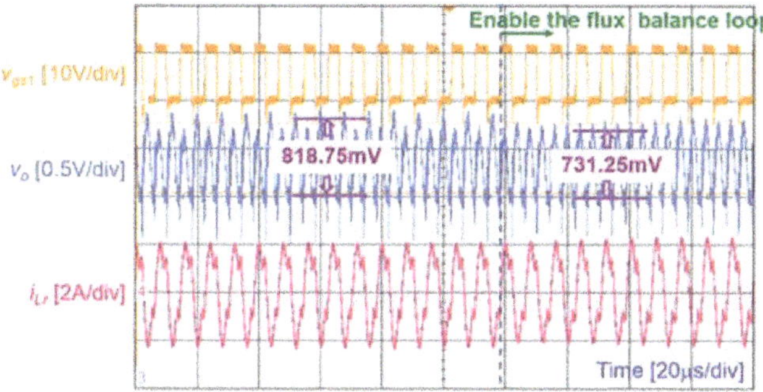

Figure 23. Experiment results of the disabled/enabled flux-balance loop at 70% load, considering the mismatched secondary side leakage inductances.

6. Conclusions

In this paper, a flux-balance loop control strategy was proposed to solve the flux walking issues of the center-tapped transformer in the LLC resonant converter, which were caused by mismatched leakage inductances at the secondary side. The magnetizing DC current effect and voltage gain parameter variation caused by the mismatched conditions at the secondary side were analyzed. Based on these analyses, the flux-balance control loop combining with the original output-voltage control loop, was proposed to resolve the issues. Besides, a magnetizing DC current sampling strategy and estimation scheme was also proposed to overcome the difficulties in measuring the magnetizing DC current. The simulation and experimental results confirmed the effectiveness of the proposed control strategy.

Author Contributions: Y.-C.L. substantially contributed to literature search, control strategy design, examination and interpretation of the results, development of the overall system, and review and proofreading of the manuscript. D.-T.C. substantially contributed to implement the control strategy, production and analysis of the results, and preparation and revision of the manuscript. C.-J.C. substantially contributed to the review and proofreading of the manuscript.

Funding: This work was supported by the Ministry of Science and Technology (MOST) in Taiwan under grants MOST 107-2221-E-002-054-MY2 and National Taiwan University under grants NTU-CC-108L890701.

Acknowledgments: The authors want to thank SIMPLIS Technologies Corporation, U.S.A, for providing SIMPLIS simulation tool.

Conflicts of Interest: The authors declare no conflict of interest.

References

1. Deng, J.; Li, S.; Hu, S.; Mi, C.; Ma, R. Design methodology of LLC resonant converters for electric vehicle battery chargers. *IEEE Trans. Veh. Technol.* **2014**, *63*, 1581–1592. [CrossRef]
2. Shen, Y.; Zhao, W.; Chen, Z.; Ca, C. Full-bridge LLC resonant converter with series-parallel connected transformers for electric vehicle on-board charger. *IEEE Access* **2018**, *6*, 13490–13500. [CrossRef]
3. Amirahmadi, A.; Domb, M.; Persson, E. High power density high efficiency wide input voltage range LLC resonant converter utilizing E-mode GaN switches. In Proceedings of the 2017 IEEE Applied Power Electronics Conference and Exposition (APEC), Tampa, FL, USA, 26–30 March 2017; pp. 350–354.
4. Joung, M.; Kim, H.; Baek, J. Dynamic analysis and optimal design of high efficiency full bridge LLC resonant converter for server power system. In Proceedings of the 2012 27th Annual IEEE Applied Power Electronics Conference and Exposition (APEC202), Orlando, FL, USA, 5–9 February 2012; pp. 1292–1297.
5. Kim, B.-C.; Park, K.-B.; Kim, C.-E.; Lee, B.-H.; Moon, G.-W. LLC resonant converter with adaptive link-voltage variation for a high-power-density adapter. *IEEE Trans. Power Electron.* **2010**, *25*, 2248–2252. [CrossRef]

6. Hu, S.; Deng, J.; Mi, C.; Zhang, M. LLC resonant converters for PHEV battery chargers. In Proceedings of the 2013 Twenty-Eighth Annual IEEE Applied Power Electronics Conference and Exposition (APEC2013), Long Beach, CA, USA, 17–21 March 2013; pp. 3051–3054.

7. Wu, H.; Mu, T.; Gao, X.; Xing, Y. A secondary-side phase-shift-controlled LLC resonant converter with reduced conduction loss at normal operation for hold-up time compensation application. *IEEE Trans. Power Electron.* **2015**, *30*, 5352–5357. [CrossRef]

8. D'Cruz, R.; Rajesh, M. Half bridge LLC resonant DC-DC converter for solar array simulator application. In Proceedings of the 2015 International Conference on Technological Advancements in Power and Energy (TAP Energy), Kollam, India, 24–26 June 2015; pp. 138–143.

9. Yang, Y.-R. A half-bridge LLC resonant converter with loose-coupling transformer and transition capacitor. In Proceedings of the 2014 9th IEEE Conference on Industrial Electronics and Applications, Hangzhou, China, 9–11 June 2014; pp. 1344–1349.

10. Huang, Y.-C.; Hsieh, Y.-C.; Lin, Y.-C.; Chiu, H.-R.; Lin, J.-Y. Study and implementation on start-up control of full-bridge LLC resonant converter. In Proceedings of the 2018 IEEE Transportation Electrification Conference and Expo, Asia-Pacific (ITEC Asia-Pacific), Bangkok, Thailand, 6–9 June 2018.

11. Lin, R.-L.; Lin, C.-W. Design criteria for resonant tank of LLC DC-DC resonant converter. In Proceedings of the 36th Annual Conference on IEEE Industrial Electronics Society (IECON2010), Glendale, AZ, USA, 7–10 November 2010; pp. 427–432.

12. Sato, M.; Nagaoka, S.; Uematsu, T.; Zaitsu, T. Mechanism of current imbalance in LLC resonant converter with center tapped transformer. In Proceedings of the 2018 International Power Electronics Conference (IPEC-Niigata 2018-ECCE Asia), Niigata, Japan, 20–24 May 2018; pp. 118–122.

13. Jung, J.-H. Bifilar winding of a center-tapped transformer including integrated resonant inductance for LLC resonant converters. *IEEE Trans. Power Electron.* **2013**, *28*, 615–620. [CrossRef]

14. Li, M.; Chen, Q.; Ren, X.; Zhang, Y.; Jin, K.; Chen, B. The integrated LLC resonant converter using center-tapped transformer for on-board EV charger. In Proceedings of the 2015 IEEE Energy Conversion Congress and Exposition (ECCE2015), Montreal, QC, Canada, 20–24 September 2015; pp. 6293–6298.

15. Panov, Y.; Jovanović, M.; Irving, B. Novel transformer-flux-balancing control of dual-active-bridge bidirectional converters. In Proceedings of the 2015 IEEE Applied Power Electronics Conference and Exposition (APEC2015), Charlotte, NC, USA, 15–19 March 2015; pp. 42–49.

16. Murakami, Y.; Sato, T.; Nishijima, K.; Nabeshima, T. Small signal analysis of LLC current resonant converters using equivalent source model. In Proceedings of the 42nd Annual Conference of the IEEE Industrial Electronics Society (IECON2016), Florence, Italy, 23–26 October 2016; pp. 1417–1422.

17. Tian, S.; Lee, F.; Li, Q. A simplified equivalent circuit model of series resonant converter. *IEEE Trans. Power Electron.* **2016**, *31*, 3922–3931. [CrossRef]

18. Tian, S.; Lee, F.; Li, Q. Equivalent circuit modeling of LLC resonant converter. In Proceedings of the 2016 IEEE Applied Power Electronics Conference and Exposition (APEC2016), Long Beach, CA, USA, 20–24 March 2016; pp. 1608–1615.

19. Fang, Z.; Wang, J.; Duan, S.; Xiao, L.; Hu, G.; Liu, Q. Rectifier current control for an LLC resonant converter based on a simplified linearized model. *Energies* **2018**, *11*, 579. [CrossRef]

20. MATLAB. System Identification Overview. Available online: https://www.mathworks.com/help/ident/gs/about-system-identification.html (accessed on 30 May 2019).

21. MATLAB. System Identification Toolbox. Available online: https://www.mathworks.com/products/sysid.html (accessed on 30 May 2019).

22. Simone, S.; Adragna, C.; Spini, C. Design guideline for magnetic integration in LLC resonant converters. In Proceedings of the 2008 International Symposium on Power Electronics, Electrical Drives, Automation and Motion (SPEEDAM2008), Ischia, Italy, 11–13 June 2008; pp. 950–957.

23. Huang, H. FHA-based voltage gain function with harmonic compensation for LLC resonant converter. In Proceedings of the 25th Annual IEEE Applied Power Electronics Conference and Exposition (APEC2010), Palm Springs, CA, USA, 21–25 February 2010; pp. 1770–1777.

Article

Circuit Topology and Small Signal Modeling of Variable Duty Cycle Controlled Three-Level LLC Converter

Hussain Humaira, Seung-Woo Baek, Hag-Wone Kim * and Kwan-Yuhl Cho

Department of Control and Instrumentation Engineering, Korea National University of Transportation, Chungju (KS001) 27469, Korea; humairabd@hotmail.com (H.H.); bsw@ut.ac.kr (S.-W.B.); kycho@ut.ac.kr (K.-Y.C.)
* Correspondence: khw@ut.ac.kr; Tel.: +82-043-841-5322

Received: 26 August 2019; Accepted: 27 September 2019; Published: 10 October 2019

Abstract: With a view to regulate the output voltage with fixed frequency, without using any additional component or complex modulation technique, a topology of a variable duty controlled three-level LLC converter is proposed and an equivalent small signal model for Electric Vehicle (EV) charger applications is deduced in this paper. The steady state equations of each operating region are derived in time domain. Based on an Extended Describing Function (EDF) approach, a small signal equivalent circuit is modeled which includes both frequency and duty controlled terms. The equivalent circuit is further simplified to derive a transfer function of duty control to output voltage. The transfer function is verified through simulation software. Analyzing the transfer function, a voltage controller is designed and implemented with a PI compensator. The simulation results of the proposed control schemes are illustrated and discussed. The topology is compared to a conventional frequency control topology and the merits of the proposed topology are presented.

Keywords: LLC Converter; Duty Control; Extended Describing Function; Small Signal Modeling

---

1. Introduction

Electricity based transportation methods are being researched all over the world for the past two decades [1] in order to find alternatives to fuel consumption [2,3]. Electric Vehicle (EV) charger systems have been attracting great attention of researchers, manufacturers and governmental agencies as concerns for energy conservation and environmental issues are growing everyday [4–6]. In the architecture of such a charger, a DC-DC converter is an essential element. From the DC link voltage supplied by the Power Factor Corrector (PFC), the DC-DC converter charges the battery through the transformer [7,8]. LLC converters have been a widespread choice for the DC-DC converter stage in EV chargers [9–12]. This type of converter is capable of achieving Zero Voltage Switching (ZVS) at the switches and Zero Current Switching (ZCS) at the rectifier diodes at almost all load conditions. It also offers a wide gain range and undergoes low conduction losses or electromagnetic interference [13–15].

An extensive amount of research has been conducted on frequency modulated LLC converter for simpler control and calculation [16–19]. However, frequency variation is conducted over a wide range resulting loss of ZVS [20] and introduces ripple [21] at certain load conditions. In [22], an additional three-level buck converter has been used to regulate the output voltage. A T-type three-level unit along with a LLC converter is applied with switch multiplexing technique [23]. The RMS value of current flowing through the secondary diodes have been reduced in [24] , however, several control modes are required to regulate the converter at rated voltage. Recently, pulse-width or duty cycle modulation are gaining interest for wide gain range [25,26], enhanced light load efficiency [27,28] and suppressed inrush current [29]. The output can be well-regulated by adding a duty cycle in the input of the resonant tank and no loss of ZVS occurs in case of duty control at any load condition. Since the switching frequency is fixed in the duty controlled converter, the magnetic components optimization is also convenient.

The reported applications of three-level LLC converters are limited in number [30–40] compared to conventional two-level converters. The voltage stress is half of the input voltage in the case of three-level converters and is advantageous for high switching frequency and higher efficiency [30]. If the switching frequency is chosen equal to resonant frequency and the ratio of resonant inductor to magnetizing inductor is smaller, voltage gain tolerance will be minimum [31]. The modulation technique in [32] with six switches is complex. Variation of output with change in frequency has been reported in [33]. The voltage is not equal to half for all loads and voltage gain in [34]. The lagging leg in [35] complicates the design and the current stress across that switch is high. Even though the converter in [36] uses fixed frequency scheme, the voltage stress in two lagging leg switches is equal to half of the input voltage. In [37], the converter is frequency controlled and calculation of additional dead time zones are required. The three-level LLC in [38] split the resonant capacitor and application of variable duty cycle is not suitable in this topology. There are eight inverter switches [39] and switching sequences can lead to complex control schemes. Nevertheless, the waveforms are highly nonlinear and the real time calculation is far more complex than frequency controlled LLC converters. A combined sliding mode control and PI control mode have been adopted in [40] for full bridge LLC converters based on an averaged large signal model. But the transfer function for PI control had to be derived through system identification method due to nonlinear behavior of the circuit . To linearize the circuit and implement a control strategy, a small signal modeling approach can be undertaken.

Although the state plane average method [41] is widely applied for PWM converters, it provides a solution up to half of the switching frequency only. Again, the sample data approach method in [42] offers a numerical way to design the compensator, which is difficult to apply. Extended Describing Function (EDF) is the preferred choice for small signal modeling approach for series/parallel resonant converters [43–45]. It is applied from the input to the output through the resonant tank. The modeling is based on first harmonic approximation of the state variables and then a set of functions is generated relating the input, state variables and control variables. The EDF method applied for LLC converter in [43,44] leads to an equivalent circuit model of fifth order. The complex terms in the functions can be withdrawn by separating the circuit into sine and cosine part and introducing cross coupled terms. In [45], the equivalent circuit is reduced to a simplified third-order circuit but for variable switching frequency control only.

In this paper, a three-level duty controlled circuit topology is proposed. The steady state waveforms are expressed by equations in time domain. The steady state behavior is observed through simulation results. With the aim of linearizing the circuit and developing a duty control to output voltage transfer function, an equivalent small signal model approach has been modeled applying the reduced third order EDF concept. Upon deriving the transfer function, it is verified by simulation. An output voltage PI controller is designed by compensating the derived transfer function. The simulation results are obtained for output by varying the reference voltage and load resistance. Lastly, a comparison between steady state wave forms of frequency control and duty control method is presented and a conclusion is drawn.

2. Duty Cycle Controlled Three-Level LLC Converter

The proposed converter consists of a dc source, two capacitors at the input, followed by the clamping diodes and a resonant tank. The rectifier along with the capacitive filter and load are separated by an isolating transformer. The circuit schematic and equivalent circuit are given in Figure 1a,b, respectively.

The input dc source voltage is split in half by C_1 and C_2. The voltage is clamped by the diodes D_1 and D_2. The half bridge configurations consist of four switches from S_1 to S_4 in series. A square voltage is generated at the input of the resonant tank, V_T, whose amplitude is equal to the half of the input dc source voltage i.e., $V_{dc}/2$. The resonant tank consists of a resonant inductor L_r, resonant capacitor C_r, and a magnetizing inductor L_m in series. The tank is followed by an isolating transformer of turns ratio n. The output side contains diode rectifiers D_3–D_6, a capacitive filter C_f and a resistive load R_L.

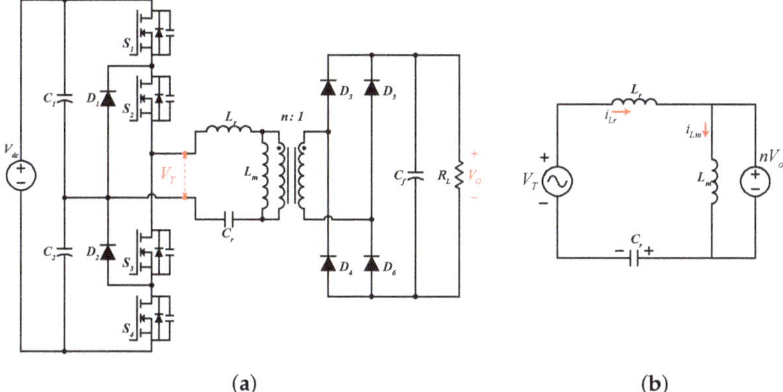

(a)   (b)

Figure 1. Proposed three-level LLC converter: (**a**) circuit schematic; and (**b**) equivalent AC circuit.

The resonant frequency f_r and the magnetizing frequency f_m of the converter are given by:

$$\omega_r = \frac{1}{\sqrt{L_r C_r}}, \quad f_r = \frac{\omega_r}{2\pi}$$
$$\omega_m = \frac{1}{\sqrt{(L_r + L_m) C_r}}, \quad f_m = \frac{\omega_m}{2\pi} \tag{1}$$

In a duty controlled converter, the switching frequency is fixed and, in the proposed converter, the switching frequency is chosen equal to the resonant frequency. The normalized parameters of the converter are given by,

$$Z_o = \sqrt{\frac{L_r}{C_r}}, \quad Z_1 = \sqrt{\frac{L_r + L_m}{C_r}}, \quad R_{ac} = \frac{8}{\pi^2} n^2 R_L, \quad m = \frac{L_m}{L_r}, \quad Q = \frac{Z_o}{R_{ac}} \tag{2}$$

where Z_o is the normalized tank impedance at resonant frequency, d Z_1 is the normalized tank impedance at magnetizing frequency, R_{ac} is the ac equivalent resistance, m is the inductance ratio and Q is the quality factor.

According to Figure 1b, generalized state space equations are given below:

$$V_{L_m}(t) = L_m \frac{di_{L_m}}{dt}(t)$$
$$i_{L_r}(t) = C_r \frac{dv_{C_r}}{dt}(t)$$
$$i_{L_m}(t) = \frac{1}{L_m} \int nV_o dt + i_{L_m}(t_0) \tag{3}$$
$$V_T(t) = L_r \frac{di_{L_r}(t)}{dt} + nV_o + V_{C_r}(t)$$

2.1. Operation of the Proposed Duty Control

The voltage and the current waveforms in the proposed three-level duty controlled topology mainly operate at six different modes. The positive half cycle is comprised of six operating regions (t_0–t_3) and the remaining three are in the negative half cycle having similar equations as the positive cycle, only opposite in polarity. The passive elements, switches and the diodes are assumed to be ideal and the short dead time is ignored. In this paper, the switches S_1 and S_4 are duty cycle controlled (≤ 0.5) and the corresponding switches S_2 and S_3 are constantly driving signals at alternating cycles. One switching period is given by $T_s = 1/f_r$. Compared to the conventional frequency controlled LLC converter, the duty controlled converter has an additional operating region, which leads to a complex analysis of the circuit parameters. The switching sequences and the key operational waveforms are depicted in Figure 2.

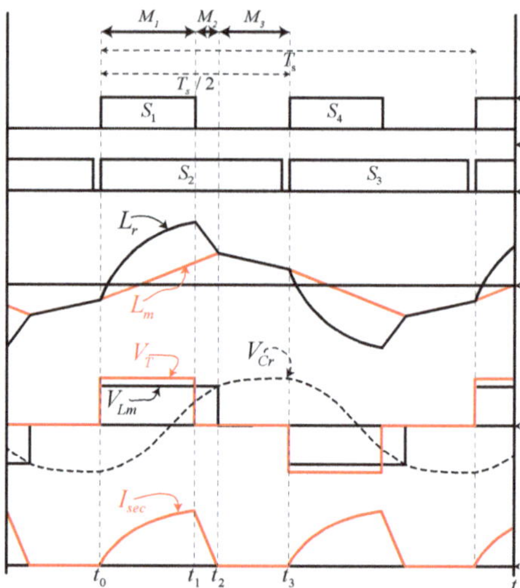

Figure 2. Operational Waveforms in duty controlled three-level LLC converter.

2.2. Steady State Operation of Operation Modes

Analyzing the state space equations and incorporating the initial conditions, the steady state equations of each operating region can be obtained. The steady state equations are discussed below and the respective equivalent circuit is presented in Figure 3a–c.

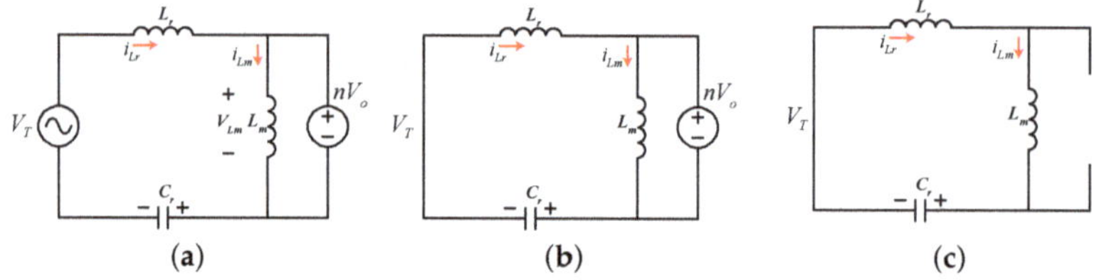

Figure 3. Equivalent steady-state circuit during: (**a**) Mode 1; (**b**) Mode 2; and (**c**) Mode 3.

Mode 1 (T_1): [From t_0 to t_1]

As shown in Figure 2, at t_0, switches S_1 and S_2 turn on. A square wave voltage V_T is produced at the input of the tank. Both I_{L_r} and I_{L_m} start linearly increasing from their negative value and reach a positive value by time t_1. The output voltage is clamped by L_m and transferred to the secondary.

The output rectifier maintains ON state. The conduction period of this mode is equivalent to $dT_s/2$. The derived steady state equations of this mode in time domain are given below:

$$I_{Lr}(t) = \frac{1}{Z_0}\left(V_T - nV_o - V_{Cr}(t_0)\right)\sin(\omega_r t) + I_{Lr}(t_0)\cos(\omega_r t)$$

$$I_{Lm}(t) = \frac{nV_o}{L_m}t + I_{Lm}(t_0) \tag{4}$$

$$V_{Cr}(t) = (V_T - nV_o) + (Z_0 I_{Lr}(t_0)\sin(\omega_r t)) - (V_T - nV_o - V_{Cr}(t_0))\cos(\omega_r t)$$

where $I_{Lr}(t_0)$, $I_{Lm}(t_0)$ and $V_{Cr}(t_0)$ are the resonant current, magnetizing current and resonant capacitor voltage at (t_0) time and also are the initial conditions of the positive half cycle. The equivalent circuit is shown in Figure 3a.

Mode 2 (T_2): [From t_1 to t_2]

The converter operation reaches its second stage at time t_1 when only switch S_2 is conducting and S_1 is turned off. Contrary to frequency controlled LLC converter where the tank input voltage remains positive in the positive half cycle, and negative in the next cycle, the square wave voltage drops to zero, consequently I_{Lr} starts to decrease. I_{Lm} is still increasing linearly. The energy difference between I_{Lr} and I_{Lm} is transferred to the secondary side. In this mode also, the output rectifiers maintain ON state. The steady state equations of this mode are the following:

$$I_{Lr}(t) = \frac{1}{Z_1}\left(-nV_o - V_{Cr}(t_1)\right)\sin(\omega_r t) + I_{Lr}(t_1)\cos(\omega_r t)$$

$$I_{Lm}(t) = \frac{nV_o}{L_m}T_2 + I_{Lm}(t_1) \tag{5}$$

$$V_{Cr}(t) = (-nV_o) + (Z_1 I_{Lr}(t_1)\sin(\omega_r t)) + (nV_o + V_{Cr}(t_1))\cos(\omega_r t)$$

Mode 3 (T_3): [From t_2 to t_3]

I_{Lr} continues to decrease and at time t_2 it is equal to I_{Lm}, which leads the converter to operate in its third operating interval. From t_2 to t_3, the resonant current I_{Lr} circulate around the tank and is equal to I_{Lm}. Both the input voltage and the output voltage are zero. The output rectifiers are in OFF state and no energy is transferred to the output. The equations expressing the behavior of the waveform in this interval are:

$$I_{Lr}(t) = -\frac{V_{Cr}}{Z_1}(t_2)\sin(\omega_m t) + I_{Lr}(t_2)\cos(\omega_m t)$$

$$I_{Lm}(t) = I_{Lr}(t) \tag{6}$$

$$V_{Cr}(t) = V_{Cr}(t_2)\cos(\omega_m t) + Z_1 I_{Lr}(t_2)\sin(\omega_m t)$$

The initial values of the operating waveforms are:

$$I_{Lm}(t_3) = -I_{Lm}(t_0)$$

$$I_{Lr}(t_3) = -I_{Lr}(t_0)$$

$$I_{Lm}(t_2) = I_{Lr}(t_2) \tag{7}$$

$$V_{Cr}(t_3) = -V_{Cr}(t_0)$$

2.3. Power Transfer Stage

The difference in energy between I_{Lr} and I_{Lm} during the time from t_0 to t_2 is transferred to the secondary side as output. The output rectifier current can be expressed as:

$$
\begin{aligned}
I_o &= \frac{2}{T_s} \int_{t_0}^{t_2} (i_{Lr}(t) - i_{Lm}(t)) dt \\
&= \frac{2}{T_s} \left[\int_{t_0}^{t_1} (i_{Lr}(t) - i_{Lm}(t)) dt + \int_{t_1}^{t_2} (i_{Lr}(t) - i_{Lm}(t)) dt \right] \\
&= \frac{2}{T_s} \left[\frac{1}{Z_o \omega_r} (V_T - nV_o - V_{Cr}(t_0))(1 - \cos \omega_r T_1) + \frac{1}{\omega_r} I_{Lr}(t_0) \sin \omega_r T_1 + \frac{nV_o}{2L_m} T_1^2 - I_{Lm}(t_0) T_1 \right. \\
&\quad \left. + \frac{1}{Z_1 \omega_r} (nV_o - V_{Cr}(t_1))(1 - \cos \omega_r T_2) + \frac{1}{\omega_r} I_{Lr}(t_1) \sin \omega_r T_2 + \frac{nV_o}{2L_m} T_2^2 - I_{Lm}(t_1) T_2 \right]
\end{aligned}
$$

$$(8)$$

As evident from Equation (8), the steady state solution is complex, non-linear and cumbersome to calculate. However, deriving the solution is not within the scope of this paper. From the next section onward, equivalent small signal modeling of the duty controlled LLC converter is discussed.

3. Equivalent Small Signal Circuit Modeling

3.1. Modeling with Extended Describing Function

In steady state operation the waveforms of the converter can be approximated as sinusoidal [41]. The resonant current, magnetizing current and the resonant voltage can be expressed in the form of their fundamental harmonic in the following manner splitting into sine and cosine parts:

$$
\begin{aligned}
i_{Lr} &= i_s(t) sin\omega_s t + i_c(t) cos\omega_s t \\
i_{Lm} &= i_{ms}(t) sin\omega_s t + i_{mc}(t) cos\omega_s t \\
V_{Cr} &= v_{Cs}(t) sin\omega_s t + v_{Cc}(t) cos\omega_s t
\end{aligned}
$$

$$(9)$$

where, ω_s is the angular switching frequency.

The derivatives of i_{Lr}, i_{Lm}, V_{Cr} are:

$$
\begin{aligned}
\frac{di_{Lr}}{dt} &= (\frac{di_s}{dt} - \omega_s i_c) sin\omega_s t + (\frac{di_c}{dt} + \omega_s i_s) cos\omega_s t \\
\frac{di_{Lm}}{dt} &= (\frac{di_{ms}}{dt} - \omega_s i_{mc}) sin\omega_s t + (\frac{di_{mc}}{dt} + \omega_s i_{ms}) cos\omega_s t \\
\frac{dV_{Cr}}{dt} &= (\frac{dv_{Cs}}{dt} - \omega_s v_{Cc}) sin\omega_s t + (\frac{dv_{Cs}}{dt} + \omega_s v_{Cc}) cos\omega_s t
\end{aligned}
$$

$$(10)$$

where, i_s, i_c, i_{ms} and i_{mc} are the sine and cosine component of the fundamental approximation of resonant current and magnetizing current respectively.

The input and the output waveforms of the inverter are approximated by the DC components whereas the resonant tank waveforms can be approximated by fundamental approximation.

$$
\begin{aligned}
V_T = v_{sq} &= f_1(v_g, d) sin\omega_s t \\
sgn(i_{Lr} - i_{Lm}) v_0 &= f_2(i_s - i_{ms}, v_{Cf}) sin\omega_s t + f_3(i_c - i_{mc}, v_{Cf}) cos\omega_s t \\
| i_{Lr} - i_{Lm} | &= f_4(i_s - i_{ms}, i_c - i_{mc}) \\
i_g &= f_5(i_{s,d})
\end{aligned}
$$

$$(11)$$

where d is the input duty cycle, v_g is the input dc voltage signal and equal to $V_{dc}/2$ in the three-level LLC Converter, i_g is the input current signal and v_0 is the output voltage signal.

The functions $f_1 \sim f_5$ are termed as EDF terms [43,44]. The EDF terms are obtained by applying Fourier transform to the nonlinear equations from Equation (3) and are shown in the following:

$$f_1(d, v_g) = \frac{4}{\pi}\sin(\frac{\pi}{2}d)v_g$$
$$f_2(i_s - i_{Lms}, v_{Cf}) = \frac{4}{\pi}\left(\frac{i_s - i_{ms}}{i_p}\right)v_{Cf}$$
$$f_3(i_c - i_{mc}, v_{Cf}) = \frac{4}{\pi}\left(\frac{i_c - i_{mc}}{i_p}\right)v_{Cf}$$
$$f_4(i_s - i_{ms}, i_c - i_{mc}) = \frac{2}{\pi}i_p \tag{12}$$
$$f_5(i_s, d) = \frac{2}{\pi}\sin(\frac{\pi}{2}d)i_s$$
$$i_p = \sqrt{(i_s - i_{ms})^2 + ((i_c - i_{mc})^2}$$

In small signal modeling, the modulation frequency is lower than the switching frequency and thus can be considered to operate in steady state condition. Substituting the Equations (10)–(12) in (3) and equating the coefficients of dc, sine and cosine terms the following harmonic balance is established

$$L_r(\frac{di_s}{dt} - \omega_s i_c) + v_{Cs} + L_m(\frac{di_{ms}}{dt} - \omega_s i_{mc}) = \frac{4}{\pi}\sin(\frac{\pi}{2}d)v_g$$
$$L_r(\frac{di_c}{dt} + \omega_s i_s) + v_{Cc} + L_m(\frac{di_{mc}}{dt} + \omega_s i_{ms}) = 0$$
$$L_m(\frac{di_{ms}}{dt} - \omega_s i_{mc}) = \frac{4}{\pi}(\frac{i_s - i_{ms}}{i_p})v_{Cf}$$
$$L_m(\frac{di_{mc}}{dt} + \omega_s i_{ms}) = \frac{4}{\pi}(\frac{i_c - i_{mc}}{i_p})v_{Cf} \tag{13}$$
$$C_r(\frac{dv_{Cs}}{dt} - \omega_s v_{Cc}) = i_s$$
$$C_r(\frac{dv_{Cc}}{dt} + \omega_s v_{Cs}) = i_c$$
$$i_g = \frac{2}{\pi}\sin(\frac{\pi}{2}d)i_s$$

Thus, the output equation can be written as:

$$v_o = \frac{2}{\pi}i_p R_L$$
$$C_f\frac{dv_{Cf}}{dt} = \frac{2}{\pi}i_p - \frac{v_{Cf}}{R_L} \tag{14}$$

When the proposed LLC converter is operating at steady state the derivative terms are considered equal to zero. In addition, when all the switches are operating at a duty cycle of 50%, the steady state solution is similar to typical two level converters and is given by:

$$M = \frac{2nV_o}{V_g} = \frac{1}{\sin(\frac{\pi}{2}\omega_n)} \left\| \frac{j\omega_n m}{j\omega_n(m+1-\frac{1}{\omega_n^2}) + \frac{\pi^2}{8}Q\frac{1}{\sin(\frac{\pi}{2}\omega_n)}(1-\omega_n^2)m\omega_n} \right\| \tag{15}$$

where ω_n is the normalized frequency given by ω_s/ω_r. The circuit is linearized and perturbed around the point $\{\hat{v}_g, \hat{d}, \hat{\omega}_s, R_L\}$ by replacing as the following,

$$v_g(t) = V_g + \hat{v}_g(t)$$
$$d(t) = D + \hat{d}(t) \tag{16}$$
$$\omega_s = \Omega_s + \hat{\omega}_s$$

where V_g, D and Ω_s are the quiescent values of the input voltage, duty and switching frequency waveforms and $\{\hat{v}_g, \hat{d}, \hat{\omega}_s\}$ are the small ac variations respectively. The EDF method is applied from the input to the output including the resonant tank. The inverter receives an input voltage from a dc source v_g and generates a high frequency square wave v_{sq}. The rectifier transforms this ac current to dc current. If switching frequency is near or equal to resonant frequency, the higher order harmonics are discarded by the resonant tank. Therefore the input and output of the converter is dominated by dc component whereas the tank is dominated by fundamental component. The average input current is directly proportional to the duty cycle d and indirectly proportional to the switching frequency since the resonant current varies the switching frequency as can be seen from Equation (13). The square wave voltage v_{sq} is perturbed to yield the duty controlled voltage source $\hat{v}_{in}$ [44],

$$\hat{v}_{in} = \frac{4}{\pi} \sin\left(\frac{D\pi}{2}\right)\Omega_s \frac{\hat{v}_g}{2} + \frac{V_g}{2}\cos\left(\frac{D\pi}{2}\right)\Omega_s \hat{d} \tag{17}$$

The inductances of the tank is also characterized by their respective complex switching frequency impedance and complex controlled voltage source due to perturbation of switching frequency. The same can be modeled for the resonant capacitor. The resonant capacitor can be replaced with an equivalent inductor when the modulation frequency is very small than the switching frequency and the current source parallel to the capacitor is replaced by an equivalent voltage source [45]. The resonant inductor, magnetizing inductor and resonant capacitor can be replaced by an equivalent series inductor L_e. Although the tank waveforms are dominated by modulation frequency which again dependent on the switching the frequency, the duty controlled input current produces a duty controlled output at the output of the resonant current. The average component of the rectified current can be expressed as the average value of the resonant current [43,44]. In [43,44], the complex terms were withdrawn by separating the sine and cosine terms and introducing cross coupled terms. The model is further reduced to a simplified third order in [45]. The equivalent inductance can be represented by L_e and can be expressed as,

$$L_e = \{L_r + L_m\left(1 - \omega_n\right)\}\,\Omega_s + \frac{1}{\Omega_s C_r} \tag{18}$$

The tank impedance at switching frequency is given by Z where Z is given by,

$$Z = \Omega_s\left\{L_r + L_m\left(1 - \omega_n\right)\right\} - \frac{1}{\Omega_s C_r} \tag{19}$$

The steady state voltage across the resonant capacitor in the sine and cosine part are given by V_{cs} and V_{cc}, and can be expressed by the resonant current divided by the capacitance, such as

$$V_{cs} = \frac{I_c}{\Omega_s C_r}, V_{cr} = \frac{I_s}{\Omega_s C_r} \tag{20}$$

The switching frequency dependent voltage source of the resonant inductor and resonant capacitor and can be combined into one source as $G_s\hat{\omega}_s$ and $G_c\hat{\omega}_s$ in the real and imaginary part respectively where,

$$G_s = L_r I_c + \frac{V_{Cs}}{\Omega_s}, G_c = L_r I_s + \frac{V_{Cc}}{\Omega_s} \tag{21}$$

The resonant current I_s and I_c are split into the magnetizing current I_{ms}, I_{mc} and the remaining flows into the output of the tank as I_{Ts} and I_{Tc} such as,

$$I_s = I_{ms} + I_{Ts}, I_c = I_{mc} + I_{Tc} \tag{22}$$

where,

$$I_{Ts} = \frac{2V_g}{\pi} \frac{\omega_n^2 L_n \left(L_n + 1 - \frac{1}{\omega_n^2} \right)}{\omega_n^2 L_n \left(L_n + 1 - \frac{1}{\omega_n^2} \right)^2 + \left(\frac{\pi^2}{8} Q \left(1 - \omega_n^2 \right) L_n \right)^2}$$

$$I_{Tc} = \frac{2V_g}{\pi} \frac{\omega_n^2 L_n^2 \left(\frac{\pi^2}{8} Q \left(1 - \omega_n^2 \right) L_n \right)}{\omega_n^2 L_n \left(L_n + 1 - \frac{1}{\omega_n^2} \right)^2 + \left(\frac{\pi^2}{8} Q \left(1 - \omega_n^2 \right) L_n \right)^2} \tag{23}$$

and,

$$I_{ms} = \frac{I_{Tc} R_{eq}}{\Omega_s L_m}, \quad I_{mc} = -\frac{I_{Ts} R_{eq}}{\Omega_s L_m} \tag{24}$$

where, $R_{eq} = R_{ac}$. The cross coupling coefficients between the output of the inverter and input of the rectifier side for the real and imaginary part can be expressed as the following,

$$k_s = \frac{2I_{Ts}}{\pi I_p}, k_c = \frac{2I_{Tc}}{\pi I_p}$$

$$k_{rs} = k_{rc} = -\frac{4}{\pi} \frac{nV_o}{I_p} \frac{I_{Ts} I_{Tc}}{I_p^2}$$

$$R_s' = \frac{4}{\pi} \frac{nV_o}{I_p} \frac{I_{Tc}^2}{I_p^2} \sin\left(\frac{\pi}{2} \omega_n \right) \tag{25}$$

$$R_c' = \frac{4}{\pi} \frac{nV_o}{I_p} \frac{I_{Ts}^2}{I_p^2} \sin\left(\frac{\pi}{2} \omega_n \right)$$

where, $I_p = \sqrt{I_{Ts}^2 + I_{Tc}^2}$.

The equivalent circuit of small signal model of three-level LLC including both frequency and duty controlled terms are given in Figure 4.

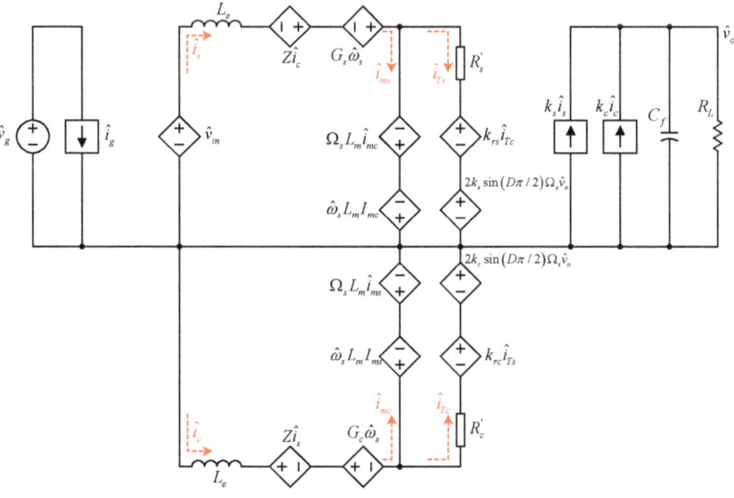

Figure 4. Equivalent small signal model for both duty and frequency controlled LLC converter.

3.2. Equivalent Circuit for Duty Control to Output

In order to derive duty control to output voltage transfer function the model is Figure 4 must be further simplified by keeping the terms containing $\hat{d}$ and $\hat{v}_o$ and discarding frequency controlled terms since the converter operates with a fixed frequency in a duty controlled converter. Again, since the fixed frequency is equal to the resonant frequency the term w_n becomes one, and from Equation (23) subsequently the parameter I_{Tc} dependent on this term becomes zero. This results into I_{ms} being zero since it is dependent on I_{Tc} as shown in Equation (24). Again, due to the same dependency on I_{Tc}, the terms $k_c, k_{rs}, k_{rc}, R'_s$ become zero from Equation (25). From Equation (15) it is apparent that at resonant frequency, the value of the magnetizing inductor, L_m doesn't affect the gain of the converter. In addition, the output current is dependent on the average value of the resonant current. In order to simplify the analysis, the controlled voltage sources associated with L_m are discarded. At resonant frequency the impedance Z can be expressed as,

$$
\begin{aligned}
Z &= \Omega_s \left\{ L_r + L_m \left(1 - w_n \right) \right\} - \frac{1}{\Omega_s C_r} \\
&= L_r - \frac{1}{C_r \Omega_s^2} \\
&= L_r \left(1 - \frac{\Omega_r^2}{\Omega_s^2} \right)
\end{aligned}
\tag{26}
$$

where, Ω_r is the resonant frequency.

At resonant frequency, the impedance Z will also become zero. Consequently, the cosine loop is decoupled and only the upper loop is taken into consideration. Therefore the model depicted in Figure 4 is further reduced as shown in Figure 5.

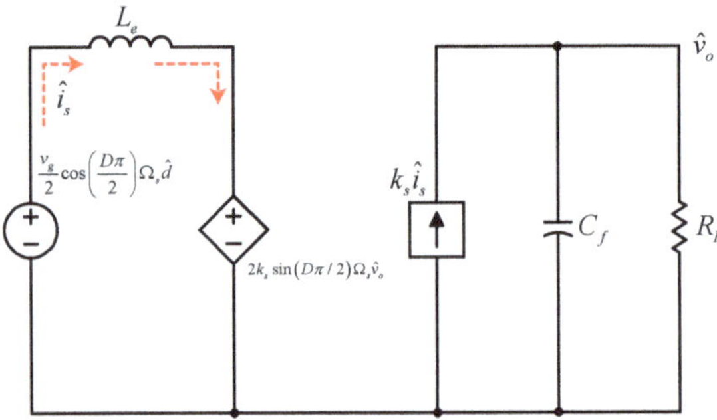

Figure 5. Simplified equivalent circuit for deriving duty control to output voltage transfer function.

Applying KVL to both the loops at the right and left, the following equations are obtained:

$$
\begin{aligned}
-\frac{V_g}{2}\cos\left(\frac{D\pi}{2}\right)\Omega_s\hat{d} + sL_e\hat{i}_s + 2k_s\sin\left(\frac{D\pi}{2}\right)\Omega_s\hat{V}_o &= 0 \\
k_s\hat{i}_s\left(\frac{sC_f R_L + 1}{sC_f}\right) &= \hat{v}_o
\end{aligned}
\tag{27}
$$

By solving the Equation (27), the transfer function of variable duty cycle control to output voltage is derived as the following:

$$
\frac{\hat{v}_o}{\hat{d}} = \frac{(s+1)\left\{ k_s\dfrac{V_g}{2}\Omega_s R_L C_f \cos\left(\dfrac{D\pi}{2}\right)\right\}}{s^2 L_e C_f + s\left\{ 2k_s^2 \sin\left(\dfrac{D\pi}{2}\right) R_L C_f \Omega_s \right\} + 2k_s^2 \sin\left(\dfrac{D\pi}{2}\right) R_L C_f \Omega_s}
\tag{28}
$$

Evidently from Equation (28), the transfer function is of second order unlike the frequency controlled converter which is of third order and the output is dependent on the values of D, Ω_s, R_L, C_f and V_g. The obtained transfer function is verified through simulation and an output voltage controller is designed shown in the following in section.

4. Simulation Verification & Controller Design

The proposed circuit schematic was simulated using the PSIM simulation software. The parameters of the converter are given in Table 1.

4.1. Simulation Results of Steady State Characteristics

The simulation results are obtained to observe the dependency of the steady state operations with variable duty and load. The simulation results at different duty cycle and load conditions are shown in Figure 6. As can be seen from Figure 6a–c the currents I_{Lr} and I_{Lm} increases linearly till t_1 and I_{Lr} decreases from that time onward. The shape of V_{Cr} is sinusoidal and reaches a positive/negative peak at the end of each positive/negative half cycle. When the duty cycle is increased in Figure 6b the interval T_2 and T_3 decreases since the interval T_1 is directly proportional to the duty cycle term and each half cycle is the sum of the three intervals. Again in Figure 6c, at low load condition, the resonant current drops to the I_{Lm} quickly compared to high load condition and circulates around the resonant tank. All the key waveforms of the converter are dependent on both duty cycle and load which increases the non-linearity of the equations. In addition, as can be seen from Figure 6, ZVS is achieved in all conditions of duty and load.

The simulation is also carried out to check the output sensitivity for resonant and magnetizing parameters. The parameters from Table 1 are increased and decreased in order to observe the change in output as a result of degradation of parameters. The duty is kept to 0.3 with respect to S_2 and S_3 and the value of load resistance is 30 ohm. When the parameters are as given in Table 1, the value of peak resonant current I_{Lr} is 11.9 A, the peak magnetizing current I_{Lm} is 1.5 A, the peak resonant capacitor voltage V_{Cr} is 113.8 V whereas the output voltage is 114.6 V. The increase in values resonant parameters will result in a lower resonant frequency and decrease in values will cause the opposite. When the resonant inductor L_r is replaced by a value of 26.565 μH and that of resonant capacitor C_r is changed to 105 nF and the magnetizing inductor is kept at 170 μH, the peak I_{Lr} is decreased to 11.1 A and the peak of V_{Cr} is decreased to 103.42 V resulting in a output voltage of 110 V. Again, when L_r is decreased to 24.12 μH and C_r is decreased to 95 nF, the output scenario is reversed. The peak I_{Lr} is increased to 12.2 A whereas the peak of V_{Cr} rises to 125.86 nF. The output voltage is increased to 119.43 V. The peak of I_{Lm} decreases to 1.45 A and increases to 1.63 A in respective cases. When only Lm is changed keeping the resonant parameters same as Table 1, the peak value of is increased to 1.6 A for the inductor value 161.5 μH and decreased to 1.4 A for for the inductor value 178 μH. However, the output voltage is not affected by the slight change in the value of magnetizing inductor. The waveforms remain undistorted in every case and is shown in Figure 7a–d respectively.

Table 1. Simulation parameters of the proposed three-level LLC Converter.

Components	V_{dc}	L_r	C_r	L_m	$f_s = f_r$	C_1, C_2	n	C_f
Value	500	25.3	100	170	100	470	1	68
Unit	V	μH	nF	μH	kHz	μF	-	μF

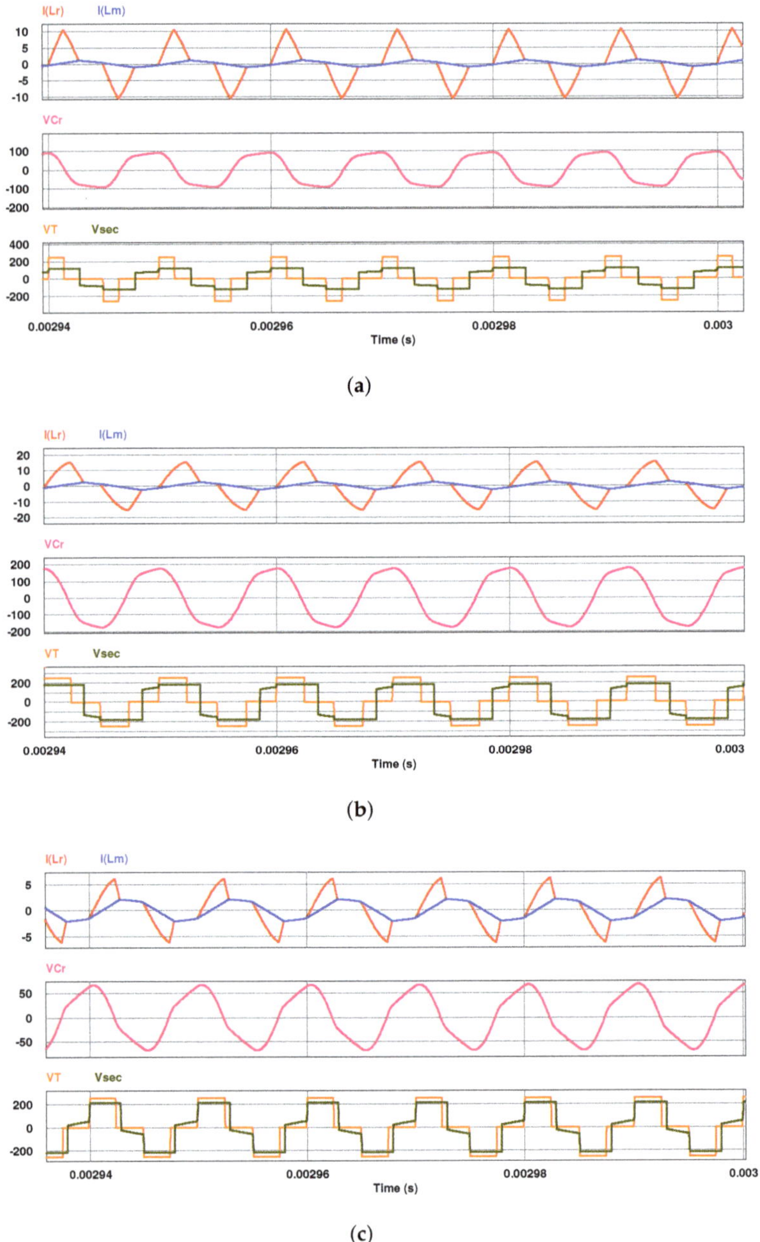

Figure 6. Simulation results for steady-state operations of three-level LLC converter operating on input voltage at 500 V: (**a**) $D = 0.3$, $R_L = 30\,\Omega$; (**b**) $D = 0.5$, $R_L = 30\,\Omega$; (**c**) $D = 0.5$, $R_L = 100\,\Omega$.

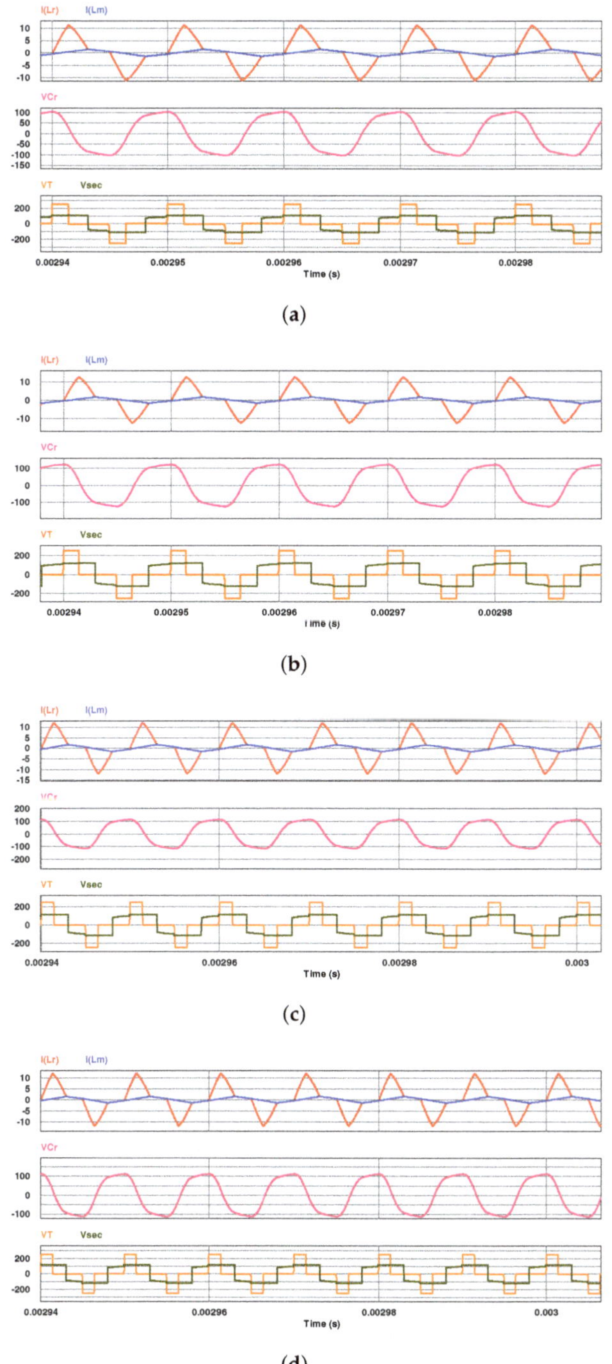

Figure 7. Simulation results for (**a**) $L_r = 26.565$ µH, $C_r = 105$ nF, $L_m = 170$ µH; (**b**) $L_r = 24.12$ µH, $C_r = 95$ nF, $L_m = 170$ µH; (**c**) $L_r = 25.3$ µH, $C_r = 100$ nF, $L_m = 161.5$ µH; (**d**) $L_r = 25.3$ µH, $C_r = 100$ nF, $L_m = 178$ µH.

4.2. Simulation Verification of Small Signal Control to Output Transfer Function & Controller Design

The transfer function obtained in Equation (28) is verified by simulation through MATLAB software with the same parameters as given in Table 1, which generates a numerical transfer function. The bode plot of the analytical and numerical transfer functions are shown in Figure 8.

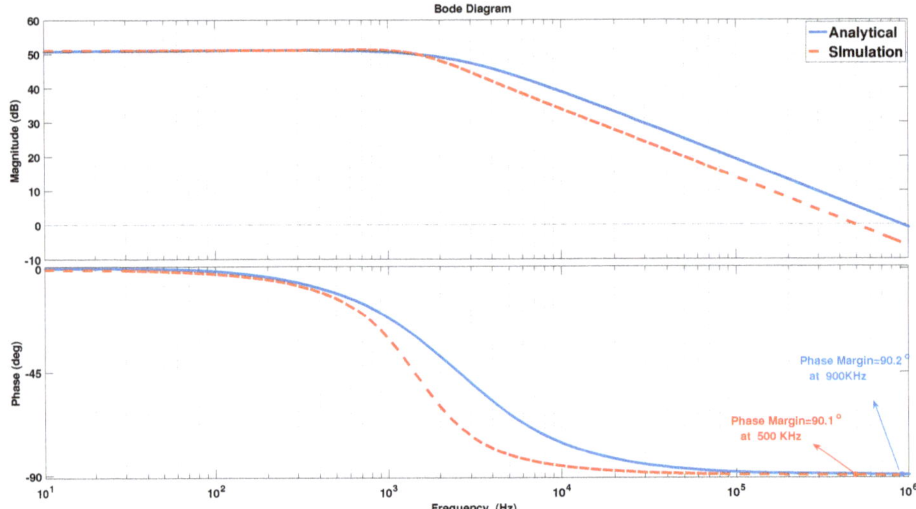

Figure 8. Verification of the analytical transfer function with MATLAB software simulation.

Figure 8 shows a good match between the obtained analytical and simulation transfer function. Both the transfer functions are of second order and contains two poles and one zero. The phase margin of the analytical transfer function is 90.2° at 90 kHz and that of the numerical transfer function is 90.1° at 50 kHz. Both the bode plots show an infinite gain margin which states that the loops are stable. However, the bandwidth is very high to implement a voltage controller. Setting the desired phase margin 70 and overshoot <10%, a PI compensator is designed by adding an extra zero to the root locus plot and further tuning. The transfer function of the compensators for the analytical and simulated transfer functions are given below:

$$C\left(s\right) = 0.000062754 \left(\frac{s + 0.0000295}{s}\right),\tag{29}$$

The uncompensated and compensated systems are plotted in Figure 9.

The bode plot of the compensated system shows a reduced bandwidth at the desired phase margin for both the analytical and simulation transfer function. The compensated transfer function is implemented as a simple PI controller transfer function to regulate the output voltage with a duty controlled signal. In a 3 level LLC converter, the expression for the gain is given by $2V_o/V_{dc}$ and since the operating frequency is equal to the resonant frequency, the gain cannot be higher than 1. Thus, the applied reference voltage must be equal to half or less than half of the input DC source voltage.

After the implementation of the PI compensator the simulation results are generated.

Firstly the controller results are verified for a fixed reference voltage. The simulation result of control signal, reference voltage, output voltage and output current is shown in Figure 10. For a 500 V input dc source, a 200 V reference was used as a reference voltage at 20 A load. The control signal is at first clamped at 0.5 which drops to 0.4 in accordance with the reference voltage.

The controller results are also checked for parameter sensitivity. Since slight variation in L_m doesn't affect the output voltage, only resonant parameters are varied and shown in Figure 11. The L_r

and C_r values are changed to 26.565 µH and 105 nF respectively for the controller results shown in Figure 11a. Again, the values are altered to 24.12 µH and 95 nF for the output presented in Figure 11b. In both the cases, the output can follow the reference.

A step value of 20 V was added to 200 V reference after 0.005 s to the converter when the input voltage source was of 500 V. The duty control signal increases and reaches to 0.4 simultaneously with the output voltage to follow the applied reference voltage of 200 V. As the step reference value is added after 0.005 s the duty controlled signal output follows the voltage reference. The duty cycle increased at 0.005 s from 0.4 to 0.44 and so does the output current.The controller can be further verified for varied load resistance for different input DC voltage for the controlled converter. The simulation results for added reference voltages and load resistances are presented in Figure 12.

Figure 12b shows the simulation output of added load resistance after 0.005 s from the beginning of simulation which results in a slight drop in the output voltage at that instant. For both Figure 12a, and Figure 12b, the output follows the reference input.

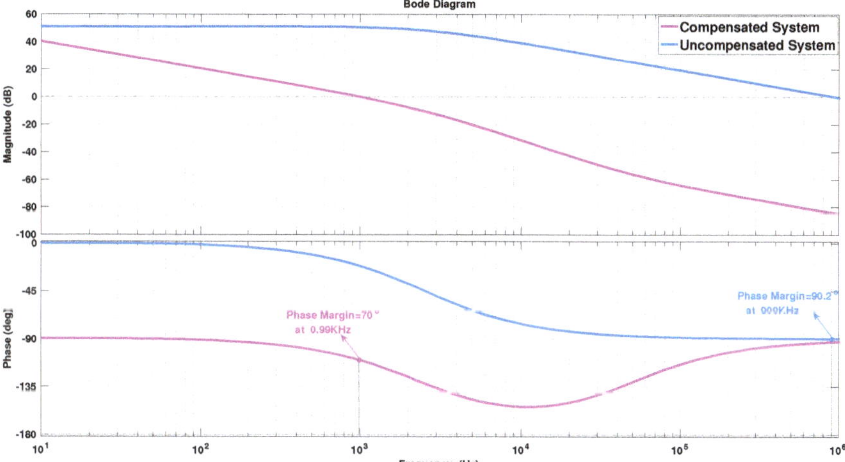

Figure 9. Bode plot of PI tuned compensated system compared to the uncompensated system.

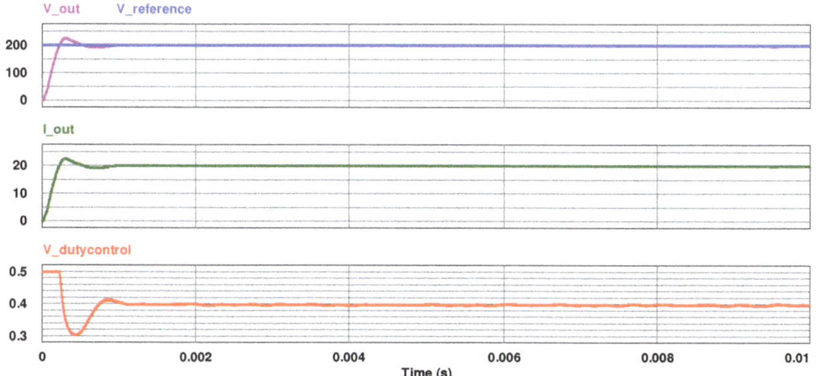

Figure 10. Simulation results for duty controlled scheme for fixed reference voltage.

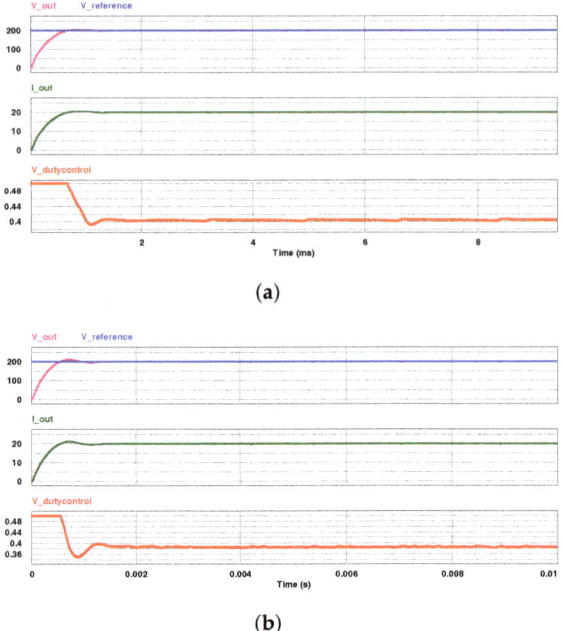

Figure 11. Controller outputs for (**a**) L_r = 26.565 µH, C_r = 105 nF, L_m = 170 µH; (**b**) L_r = 24.12 µH, C_r = 95 nF, L_m = 170 µH.

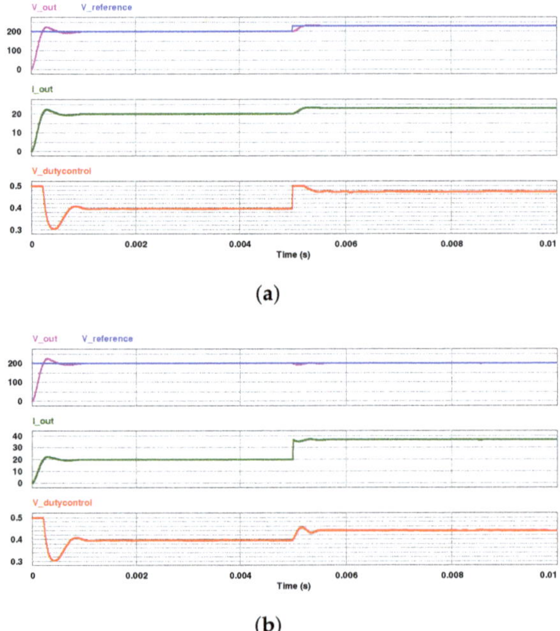

Figure 12. Controller outputs for (**a**) Varied reference voltage; (**b**) Varied load resistance.

The reduced small signal model is of second order, unlike the three order model for switching frequency control to output voltage and can be implemented with a simple PI compensator. To further validate the simplicity of the voltage regulation through duty control, the steady state waveforms are compared with the frequency controlled waveforms. The parameters of the converter are the same as listed in Table 1 and the value of the load resistance is kept at 50 Ω. In order to obtain a desired output voltage of 120 V, input and output signals of both duty control and frequency control method in steady state are presented in Figure 13.

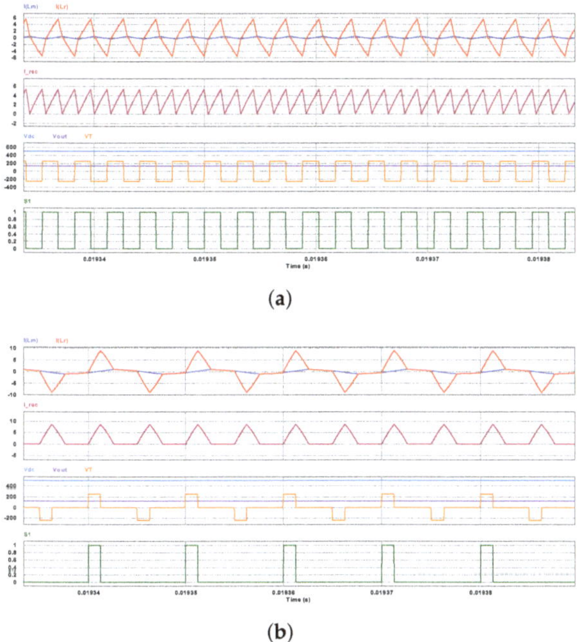

(a)

(b)

Figure 13. Steady state waveforms of (**a**) Frequency controlled; (**b**) Duty controlled three-level LLC Converter.

Figure 13a shows the converter waveforms when the switching frequency is varied. To achieve a voltage as low as 120 V, the converter has to operate at a very high switching frequency. In this case, the switching frequency is required to regulate the voltage at 120 V is 340 K which is 3.4 times the resonant frequency of the tank. The high switching frequency can cause switching loss and decrease in efficiency. Moreover, the parameters of the converter should also be should be suitably optimized for frequency control for operating at wide range of frequencies. Further control methods have to be applied to obtain the desired output voltage at a reduced switching frequency. As the lighter load condition is approached, the required switching frequency becomes higher which ultimately results in large switching loss without ZCS of the rectifier diodes $D_3 \sim D_6$. However, the duty control method is rather simple to implement and the desired output voltage can be achieved. The waveforms of duty controlled converter are presented in Figure 13b. In this case, the duty cycle value is tuned at 0.27 with respect to the switches S_2 and S_4. If the load is further decreased, the applied duty cycle must be of greater value to get a lower output voltage than $V_{dc}/2$. Although both the duty control and frequency controlled outputs are dependent on load condition, the desired voltage for duty controlled converter can be achieved at the fixed switching frequency, only by adjusting the duty cycle. Thus the control is much simpler and causes minimum switching loss. The soft switching conditions are maintained at almost all load conditions. The optimization of parameters of the converter is not also as crucial as for the frequency controlled three-level LLC. converter. Therefore, the proposed topology can be regarded as a superior method in terms of simple and convenient control.

5. Conclusions

The proposed circuit topology of three-level duty controlled LLC converter is presented in this paper for a potential EV charger application. The modulation technique is rather simple and no additional component or circuitry has been required for controlling the output. The operating regions are discussed and the steady state equations of the respective regions are derived. The steady state waveforms are observed through simulation for different duty cycle and load resistance values. Using the EDF concept, a linearized small signal equivalent circuit is modeled for duty control to output voltage and a reduced second-order transfer function is derived. The transfer function is verified through simulation and compensated to design a voltage controller. The controller outputs are obtained for varying step reference voltage and load resistance to validate the controller design. Finally, the steady state conditions between the variable frequency control and variable duty cycle control are presented. Contrary to the frequency control method, the voltage can be regulated by changing the duty cycle only while the converter operates at the resonant frequency, making the proposed topology much simpler to implement.

Author Contributions: H.H. derived the model, generated the simulation results and prepared the original draft. S.-W.B. assisted in implementation of results and preparing the draft. H.-W.K. proposed the idea and supervised the paper. K.-Y.C. contributed in idea development and discussion of the results.

Funding: This work was supported by the Industrial Strategic Technology Development Program (KEIT) funded By the Ministry of Trade, Industry & Energy (MOTIE) of the Republic of Korea (No. 201801180001). This work was supported by "Human Resources Program in Energy Technology" of the Korea Institute of Energy Technology Evaluation and Planning (KETEP), granted financial resource from the Ministry of Trade, Industry & Energy, Republic of Korea (No. 20184030202270).

Conflicts of Interest: The authors declare no conflict of interest.

Abbreviations

The following abbreviations are used in this manuscript:

V_{dc}	Input dc source voltage
V_T	Primary side voltage
C_1, C_2	Input side capacitors
D_1, D_2	Clamping diodes at the input
$D_3 \sim D_6$	Rectifier diodes
$S_1 \sim S_4$	Switches
ω_r	Resonant angular frequency
ω_s	Switching angular frequency
ω_m	Angular frequency due to magnetizing inductor
f_r	Resonant frequency
f_m	Frequency due to magnetizing inductor
n	Turns ratio
f_s	Switching frequency
d	Duty cycle
t	time
M	Voltage gain
L_r	Resonant inductor
C_r	Resonant capacitor
L_m	Magnetizing inductor
L_m	Magnetizing inductor
C_f	Output capacitor

R_L	Load resistance
T_s	Switching period
Z_0	Normalized tank impedance at resonant frequency
Z_1	Normalized tank impedance at magnetizing frequency
R_{ac}	ac equivalent resistance
m	L_m / L_r
Q	Quality factor
I_{L_r}	Resonant inductor current
I_{L_m}	Magnetizing inductor current
V_{C_r}	Voltage across resonant capacitor
V_{L_m}	Voltage across magnetizing inductor
V_o	Output voltage
I_o	Output current
V_{sec}	Secondary side voltage

References

1. Kyriakopoulos, G.L.; Arabatzis, G. Electrical energy storage systems in electricity generation: Energy policies, innovative technologies, and regulatory regimes. *Renew. Sustain. Energy Rev.* **2016**, *56*, 1044–1067. [CrossRef]
2. Kyriakopoulos, G.L.; Arabatzis, G.; Tsialis, P.; Ioannou, K. Electricity consumption and RES plants in Greece: Typologies of regional units. *Renew. Energy* **2018**, *127*, 134–144. [CrossRef]
3. Arabatzis, G.; Kyriakopoulos, G.; Tsialis, P. Typology of regional units based on RES plants: The case of Greece. *Renew. Sustain. Energy Rev.* **2017**, *78*, 1424–1434. [CrossRef]
4. Chakraborty, S.; Vu, H.-N.; Hasan, M.M.; Tran, D.-D.; Baghdadi, M.E.; Hegazy, O. DC-DC Converter Topologies for Electric Vehicles, Plug-in Hybrid Electric Vehicles and Fast Charging Stations: State of the Art and Future Trends. *Energies* **2019**, *12*, 1569. [CrossRef]
5. Rezvani, Z.; Jansson, J.; Bodin, J. Advances in consumer electric vehicle adoption research: A review and research agenda. *Transp. Res. Part D Transp. Environ.* **2015**, *34*, 122–136. [CrossRef]
6. Helmers, E.; Marx, P. Electric cars: Technical characteristics and environmental impacts. *Environ. Sci. Eur.* **2012**, *24*, 14. [CrossRef]
7. Asa, E.; Colak, K.; Bojarski, M.; Czarkowski, D. Three phase LLC resonant converter with D-DLL control technique for EV battery chargers. In Proceedings of the 2014 IEEE International Electric Vehicle Conference (IEVC), Florence, Italy, 17–19 December 2014.
8. Gautam, D.S.; Musavi, F.; Eberle, W.; Dunford, W.G. A Zero-Voltage Switching Full-Bridge DC–DC Converter With Capacitive Output Filter for Plug-In Hybrid Electric. *IEEE Trans. Power Electron.* **2013**, *28*, 5728–5735. [CrossRef]
9. Yang, B.; Lee, F.C.; Zhang, A.J.; Huang, G. LLC resonant converter for front end DC/DC conversion. In Proceedings of the Seventeenth Annual IEEE Applied Power Electronics Conference and Exposition, Dallas, TX, USA, 10–14 March 2002; pp. 1108–1112.
10. Musavi, F.; Craciun, M.; Gautam, D.S.; Eberle, W.; Dunford, W.G. An LLC Resonant DC–DC Converter for Wide Output Voltage Range Battery Charging Applications. *IEEE Trans. Power Electron.* **2013**, *28*, 5437–5445. [CrossRef]
11. Shafiei, N.; Arefifar, S.A.; Saket, M.A.; Ordonez, M. High efficiency LLC converter design for universal battery chargers. In Proceedings of 2016 IEEE Applied Power Electronics Conference and Exposition (APEC), Long Beach, CA, USA, 20–24 March 2016; pp. 2561–2566.
12. Luo, J.; Wang, J.; Fang, Z.; Shao, J.; Li, J. Optimal Design of a High Efficiency LLC Resonant Converter with a Narrow Frequency Range for Voltage Regulation. *Energies* **2018**, *11*, 1124. [CrossRef]
13. Hu, H.; Fang, X.; Chen, F.; Shen, Z.J.; Batarseh, I. A Modified High-Efficiency LLC Converter With Two Transformers for Wide Input-Voltage Range Applications. *IEEE Trans. Power Electron.* **2013**, *28*, 1946–1960. [CrossRef]
14. Musavi, F.; Craciun, M.; Gautam, D.S.; Eberle, W. Control Strategies for Wide Output Voltage Range LLC Resonant DC–DC Converters in Battery Chargers. *IEEE Trans. Veh. Technol.* **2014**, *63*, 1117–1125. [CrossRef]
15. Deng, J.; Li, S.; Hu, S.; Mi, C.C.; Ma, R. Design Methodology of LLC Resonant Converters for Electric Vehicle Battery Chargers. *IEEE Trans. Veh. Technol.* **2014**, *63*, 1581–1592. [CrossRef]

16. Feng, W.; Lee, F.C.; Mattavelli, P. Simplified Optimal Trajectory Control (SOTC) for LLC Resonant Converters. *IEEE Trans. Power Electron.* **2013**, *28*, 2415–2426. [CrossRef]
17. Kim, D.-H.; Kim, M.-S.; Hussain Nengroo, S.; Kim, C.-H.; Kim, H.-J. LLC Resonant Converter for LEV (Light Electric Vehicle) Fast Chargers. *Electronics* **2019**, *8*, 362. [CrossRef]
18. Fang, Z.; Wang, J.; Duan, S.; Shao, J.; Hu, G. Stability Analysis and Trigger Control of LLC Resonant Converter for a Wide Operational Range. *Energies* **2017**, *10*, 1448. [CrossRef]
19. Khalid, U.; Khan, M.M.; Khan, M.Z.; Rasool, U.; Ahmad, M.; Xu, J. Pulse Width and Frequency Hybrid Modulated LLC Converter Adapted to Ultra Wide Voltage Range. *Inventions* **2018**, *3*, 77. [CrossRef]
20. Feng, W.; Lee, F.C.; Mattavelli, P.; Huang, D.; Prasantanakorn, C. LLC resonant converter burst mode control with constant burst time and optimal switching pattern. In Proceedings of the 2011 Twenty-Sixth Annual IEEE Applied Power Electronics Conference and Exposition (APEC), Fort Worth, TX, USA, 6–11 March 2011; pp. 6–12.
21. He, P.; Khaligh, A. Comprehensive Analyses and Comparison of 1 kW Isolated DC–DC Converters for Bidirectional EV Charging Systems. *IEEE Trans. Transp. Electrif.* **2017**, *3*, 147–156. [CrossRef]
22. Cai, G.; Liu, D.; Liu, C.; Li, W.; Sun, J. A High-Frequency Isolation (HFI) Charging DC Port Combining a Front-End Three-Level Converter with a Back-End LLC Resonant Converter. *Energies* **2017**, *10*, 1462.
23. Gao, Y.; Cai, W.; Yi, F. A single-stage single-phase isolated AC-DC converter based on LLC resonant unit and T-type three-level unit for battery charging applications. In Proceedings of the 2016 IEEE Applied Power Electronics Conference and Exposition (APEC), Long Beach, CA, USA, 20–24 March 2016; pp. 1861–1867.
24. Yan Li, Kun Zhang, and Shuaifei Yang Multimode Hybrid Control Strategy of LLC Resonant Converter in Applications with Wide Input Voltage Range. *J. Power Electron.* **2019**, *19*, 201–210.
25. Yang, Z.; Wang, J.; Ma, H.; Du, J. A wide output voltage LLC Series Resonant Converter with hybrid mode control method. In Proceedings of the 2015 IEEE 2nd International Future Energy Electronics Conference (IFEEC), Taipei, Taiwan, 1–4 November 2015.
26. Sun, X.; Shen, Y.; Zhu, Y.; Guo, X. Interleaved Boost-Integrated LLC Resonant Converter with Fixed-Frequency PWM Control for Renewable Energy Generation Applications. *IEEE Trans. Power Electron.* **2015**, *30*, 4312–4326. [CrossRef]
27. Lo, Y.; Lin, C.; Hsieh, M.; Lin, C. Phase-Shifted Full-Bridge Series-Resonant DC-DC Converters for Wide Load Variations. *IEEE Trans. Power Electron.* **2011**, *58*, 2572–2575. [CrossRef]
28. Kim, J.; Kim, C.; Kim, J.; Lee, J.; Moon, G. Analysis on Load-Adaptive Phase-Shift Control for High Efficiency Full-Bridge LLC Resonant Converter Under Light-Load Conditions. *IEEE Trans. Power Electron.* **2016**, *31*, 4942–4955.
29. Yang, D.; Chen, C.; Duan, S.; Cai, J.; Xiao, L. A Variable Duty Cycle Soft Startup Strategy for LLC Series Resonant Converter Based on Optimal Current-Limiting Curve. *IEEE Trans. Power Electron.* **2016**, *31*, 7996–8006. [CrossRef]
30. Chen, G.; Li, H.; Sun, X.; Wang, Y. A Method for Designing Resonant Tank of Half-Bridge Three-Level LLC Converter. In Proceedings of the 2018 Chinese Automation Congress (CAC), Xi'an, China, 30 November–2 December 2018; pp. 2312–2316.
31. Haga, H.; Kurokawa, F. Modulation Method of a Full-Bridge Three-Level LLC Resonant Converter for Battery Charger of Electrical Vehicles. *IEEE Trans. Power Electron.* **2017**, *32*, 2498–2507. [CrossRef]
32. Haga, H.; Maruta, H.; Kurokawa, F. A comparative study of voltage gain tolerance in conventional and three-level LLC converters against circuit variation. In Proceedings of the 2016 IEEE International Conference on Renewable Energy Research and Applications (ICRERA), Birmingham, UK, 20–23 November 2016; pp. 153–157.
33. Haga, H.; Kurokawa, F. Dynamic analysis of the three-level LLC resonant converter for a rectifier in HVDC distribution system. In Proceedings of the 2015 IEEE International Telecommunications Energy Conference (INTELEC), Osaka, Japan, 18–22 October 2015.
34. Jiang, T.; Zhang, J.; Wu, X.; Sheng, K.; Wang, Y. A Bidirectional Three-Level LLC Resonant Converter with PWAM Control. *IEEE Trans. Power Electron.* **2016**, *31*, 2213–2225. [CrossRef]
35. Guo, Z.; Sha, D.; Liao, X. Hybrid Phase-Shift-Controlled Three-Level and LLC DC–DC Converter with Active Connection at the Secondary Side. *IEEE Trans. Power Electron.* **2015**, *30*, 2985–2996. [CrossRef]
36. Jin, K.; Ruan, X. Hybrid Full-Bridge Three-Level LLC Resonant Converter- A Novel DC-DC Converter Suitable for Fuel Cell Power System. In Proceedings of the 2005 IEEE 36th Power Electronics Specialists Conference, Recife, Brazil, 12–16 June 2005; pp. 361–367.

37. Gu, Y.; Lu, Z.; Hang, L.; Qian, Z.; Huang, G. Three-level LLC series resonant DC/DC converter. *IEEE Trans. Power Electron.* **2005**, *20*, 781–789. [CrossRef]
38. Lee, I.; Moon, G. Analysis and Design of a Three-Level LLC Series Resonant Converter for High- and Wide-Input-Voltage Applications. *IEEE Trans. Power Electron.* **2012**, *27*, 2966–2979. [CrossRef]
39. Canales, F.; Li, T.H.; Aggeler, D. Novel modulation method of a three-level isolated full-bridge LLC resonant DC-DC converter for wide-output voltage application. In Proceedings of the 2012 15th International Power Electronics and Motion Control Conference (EPE/PEMC), Novi Sad, Serbia, 4–6 September 2012.
40. Zheng, K.; Zhang, G.; Zhou, D.; Li, J.; Yin, S. Modeling, Dynamic Analysis and Control Design of Full-Bridge LLC Resonant Converters with Sliding-Mode and PI Control Scheme. *J. Power Electron.* **2018**, *18*, 766–777.
41. Davoudi, A.; Jatskevich, J.; Rybel, T.D. Numerical state-space average-value modeling of PWM DC-DC converters operating in DCM and CCM. *IEEE Trans. Power Electron.* **2006**, *21*, 1003–1012. [CrossRef]
42. Subhash Joshi, T.G.; John, V. Small signal audio susceptibility model for Series Resonant Converter. In Proceedings of the 2016 IEEE International Conference on Power Electronics, Drives and Energy Systems (PEDES), Trivandrum, India, 14–17 December 2016.
43. Yang, E.X.-Q. Extended Describing Function Method for Small-Signal Modeling of Resonant and Multi-Resonant Converters. Ph.D. Thesis, Virginia Polytechnic Institute and State University, Blacksburg, VA, USA, February 1994.
44. Yang, E.X.; Lee, F.C.; Jovanovic, M.M. Small-signal modeling of series and parallel resonant converters. In Proceedings of the APEC'92 Seventh Annual Applied Power Electronics Conference and Exposition, Boston, MA, USA, 23–27 Feburary 1992; pp. 785–792.
45. Tian, S.; Lee, F.C.; Li, Q. A Simplified Equivalent Circuit Model of Series Resonant Converter. *IEEE Trans. Power Electron.* **2016**, *31*, 3922–3931. [CrossRef]

Article

Design Considerations of Series-Connected Devices Based LLC Converter

Dongkwan Yoon, Sungmin Lee and Younghoon Cho *

Department of Electrical Engineering, Konkuk University, Seoul 05029, Korea; asdf0462@konkuk.ac.kr (D.Y.); lsm2429@konkuk.ac.kr (S.L.)
* Correspondence: yhcho98@konkuk.ac.kr; Tel.: +82-10-6207-0431

Received: 30 November 2019; Accepted: 2 January 2020; Published: 5 January 2020

Abstract: This paper describes the design of a Series-Connected Device based on a fixed–frequency LLC resonant converter (SCDLLC). Isolation of the dc-dc converter like the LLC resonant converter is used for the stability of the high voltage system such as a solid-state-transformer (SST). The series-connected devices driving method is one of the methods applicable to a high voltage system. When driving series-connected devices, an auxiliary circuit for voltage balancing between series-connected devices is required, which can be simply implemented using a passive element. In this paper, LLC converter design with balancing circuits configured in parallel with a device is provided, and both the simulations and experiments were performed.

Keywords: solid-state-transformer (SST); isolation dc-dc converter; LLC resonant converter; series-connected devices

---

1. Introduction

Recently, with new technologies such as smart grid and DC distribution, SST have emerged. SST connects directly to the grid instead of the traditional transformers to perform a variety of roles such as power factor correction and DC distribution. Therefore, SST should be able to cope with the high voltage of the grid, so that the various studies can be conducted [1–8]. To cope with high voltage, a multi-module converter (MMC) or a series-connected switching devices method has been studied. Figure 1 shows the difference between MMC and the series-connected device method.

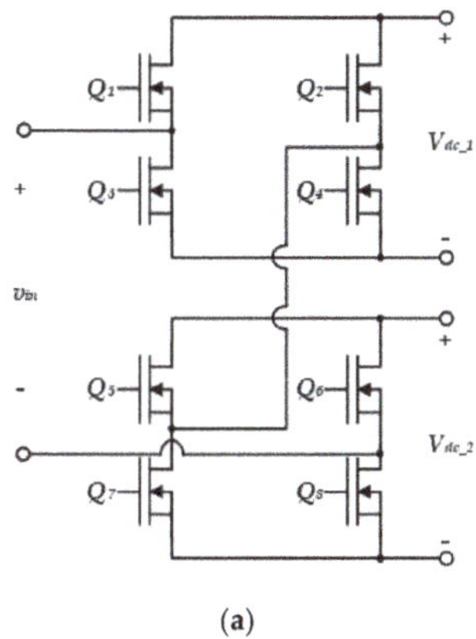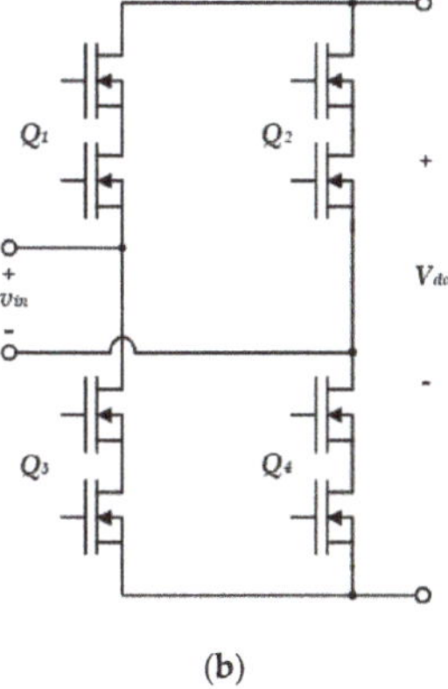

(a)   (b)

Figure 1. Comparison structure between MMC and the series-connected devices-based converter. (**a**) MMC; (**b**) series-connected devices-based converter.

MMC is a structure that increases the front of the circuit by stacking the circuit in series, which is relatively easy to expand, and has the advantage of reducing filter size by increasing the Power Conversion System (PCS) voltage level. But, since there is a separated DC link voltage, it is necessary to control and balance the separated DC link voltage. In addition, all individual devices should be controlled as the number of modules increases, which requires a digital computing device with many Pulse Width Modulation (PWM) channels. On the other hand, the series-connected devices behave like a single device, so even if many devices are connected in series, no digital computing device with many PWM channels is required. Furthermore, due to the series-connected devices, PCS does not have the separated DC link voltage. No additional control method is needed for DC link voltage balancing. On the other hand, because the output voltage level of series-connected devices based on PCS decreases, the filter size is larger than the MMC method. In addition, there is an issue with a switch voltage imbalance between the series-connected switches.

According to References [9–16], there are various reasons for causing a voltage imbalance. The first reason is the error of the gate driver. The gate driver is composed of various passive components and semiconductor devices like Negative, Positive, Negative (NPN) and Positive, Negative, Positive (PNP) transistors. Errors in devices of the gate driver cause an unbalance in the gate signal, which causes an imbalance in the voltage across the series-connected devices. The second reason is the error of parasitic components of the device itself. The error of the output capacitor of devices affects the switching speed and causes a voltage imbalance. The third reason is the parasitic capacitor from gate to ground of each device.

To solve voltage imbalance, various studies have been conducted [9–16]. In Reference [9], Active Gate Driver using Field Programmable Gate Array (FPGA) with RCD (Resistor, Capacitor and Diode) snubbers across each device method for balancing was proposed and tested. In the documents, the three-phase half-bridge inverter was built in 12-series connected Insulated Gage Bipolar Transistor (IGBT) and verified at a rated current within ±10 kV DC-link. In Reference [12], an active gate driver using a current mirror with a steady-state balancing resistor was proposed and performed. The steady-state voltage imbalance is reduced by the balancing resistor, and the transient imbalance was solved by the analog gate controller used a current mirror. In Reference [14], the voltage balancing method with only passive components using a single gate driver unit was proposed. In the documents, experimental verifications were applied to the DC circuit breaker with a 1.2 kV bus voltage. It was confirmed that it is possible to cope with the high voltage system by using a series connection element applied to various methods.

In a high voltage system, like SST, isolated converters are needed to increase the stability of the system and grid. Isolated converters are one of the necessary converters for SST regardless of MMC or series-connected devices. There are many types of isolated converters such as Dual-Active-Bridge (DAB), Quad-Active-Bridge (QAB), and LLC Converter. A fixed-frequency resonant converter conducts zero voltage switching (ZVS) over a wide frequency range and provides the advantages of high efficiency and high-power density.

In this paper, fixed-frequency LLC resonant converter-based series-connected devices is proposed. In order to cope with a high voltage, a series-connected device based on PCS was applied and an R-C snubber was applied to solve the voltage imbalance between series-connected devices. In conclusion, this paper examined the effect of R-C snubber applied to the LLC converter to solve the voltage imbalance. The proposed converter consists of a full-bridge inverter. Unlike the usual full-bridge converter, the proposed converter consists of eight devices, as shown in Figure 1b. The operation of the SCDLLC converter with a balancing circuit is described. This paper is structured as follows. In Section 2, imbalance factors of series-connected devices are described and a balancing method using passive elements is expressed. In Section 3, a configuration method of the LLC converter considering series-connected devices is explained. In Sections 4 and 5, a 3-kW prototype model is manufactured to verify the usefulness of the proposed system and the simulations and experiments are performed. Lastly, in Section 6, we discuss experimental results and conclude the paper.

2. Design of Series-Connected Devices Balancing Circuit

Figure 2 depicts the series-connected device with various factors that cause a voltage imbalance. Gate signal mismatch can occur due to an error of the device's parasitic capacitor, unbalance of the gate signal due to a gate pattern, and an error of the parasitic capacitor from gate to ground, which causes an unbalance of voltage across the device. Furthermore, the voltage imbalance of devices increases the stress and can cause damage to the device. In this section, in order to prevent damage to the device, the voltage balancing method of the series-connected devices was analyzed using passive components.

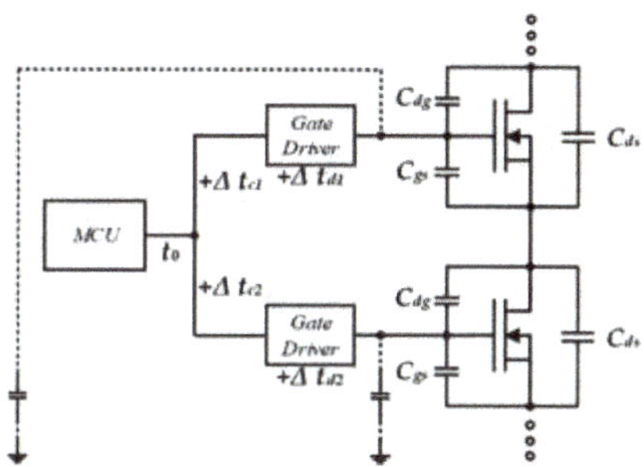

Figure 2. Voltage imbalance factors in series-connected devices.

Passive snubbers implemented by passive devices (C, RC, RCD, etc.) are used for voltage balancing circuits. Voltage imbalance of devices is somewhat limited by placing the passive snubber circuit between drain to source. The use of passive snubber circuits reduces the switching speed of devices and the voltage imbalance due to a gate signal imbalance. Figure 3 shows series-connected devices with passive balancing circuits. The snubber circuit composed of a steady-state balancing resistor (R_{bal}) and a transient state balancing capacitor (C_{snub}, R_{damp}).

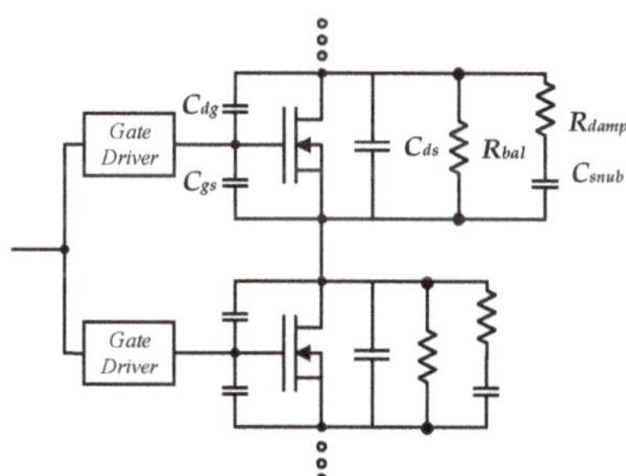

Figure 3. Structure of voltage balancing circuits in series-connected devices.

The steady-state balancing resistors operate when the switch is fully turned-off. If the steady-state balancing resistors are small, the devices are well balanced. However, the power dissipation of balancing resistors increased, which requires the use of higher-rated resistors. Conversely, If the balancing resistors are big, the power dissipation of balancing resistors decrease. However, there is a high probability that voltage balancing between series-connected devices will fail. Therefore, when designing the balancing resistor, the leakage current of the device should be considered. Steady-state resistance are calculated below.

$$\frac{v_{ds}}{10 \times i_{leakage\_max}} < R_{bal} < \frac{v_{ds}}{10 \times i_{leakage\_min}} \tag{1}$$

The transient state balancing circuit (C_{snub}, R_{damp}) operates when the switch is a turn-on state and a turn-off state. C_{snub} is selected considering the parasitic capacitor of the device. To avoid imbalance by parasitic capacitor errors of the devices, attach a balancing capacitor with negligible parasitic capacitors of a device. If snubber capacitance is big, voltage unbalance during the transient will be stabilized, but switching speed will decrease and switching loss will increase. If the snubber capacitance is small, the snubber capacitors cannot perform proper balancing. R_{snub} works as a damping resistor. The inrush current of the snubber capacitor at the switching state is limited by R_{snub}. Therefore, R_{snub} is selected by the current rating of the snubber capacitor. In addition, when calculating the R_{snub} value, a time constant ($\tau = RC$) and switching frequency (f_{sw}) should be considered. A large time constant reduces the switching speed, increases switching losses, and also hinders soft switching of the LLC converter.

3. Design of LLC Converter

Figure 4 depicts the series-connected SiC MOSFETs LLC Converter with voltage balancing circuits. Generally, the output capacitance of the switching device should be considered in the design process. However, in this case, Snubber capacitors are added in parallel with the series-connected devices for voltage balancing. The output capacitor should be discharged during deadtime for a sufficient Zero Voltage Switching (ZVS) effect. Therefore, the snubber capacitor should be considered when designing an LLC converter. Except for the snubber capacitor, the others process of the designed LLC converter are followed by the general LLC converter design process in References [17,18].

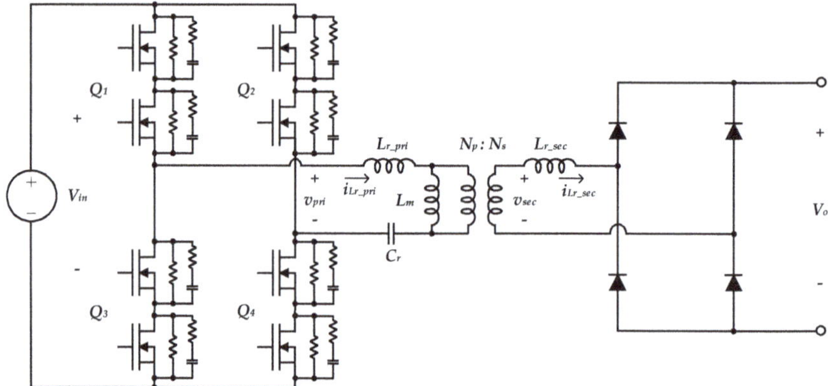

Figure 4. Series-connected devices-based LLC Converter with a voltage balancing circuit.

Figure 5 shows theoretical waveforms of the conventional LLC converter. For simplifying it to analyze, some assumptions are made as follows.

- There is no error of parasitic elements of a switching device;
- Gate drivers are all ideal, and there is no time delay between series-connected devices due to the gate driver.

The gate signals of devices are off during the deadtime (t_{dead}) interval. The magnetizing current and resonant current become the same during the deadtime. During the deadtime, inductor current flows through the antiparallel diode of the switch and discharges the parasitic output capacitor of the switch. Then, the switch turns on after zero voltage is formed by the discharged output capacitor so that soft switching is possible. The equation of minimum deadtime and magnetizing inductance for soft switching is calculated below.

$$t_{dead} \geq 16 C_{oss} f_{sw} L_m \tag{2}$$

Equation (2) represents that the ZVS is related to the magnetizing inductance of the high frequency transformer (HF transformer) and the output capacitance of the switch. As mentioned in the previous section, Snubber capacitors are required for voltage balance between series-connected devices. Therefore, when manufacturing the LLC converter with balancing circuits that are connected in parallel to each device, balancing circuits should be considered. This is not a parasitic component of the switch. After that, the rest process design of an LLC converter follows the conventional design process.

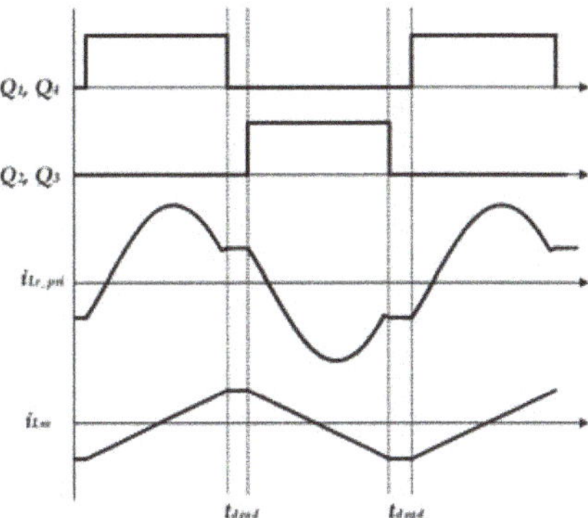

Figure 5. Switching state and resonant (i_{Lr_pri}), magnetizing current (i_{Lm}) of the transformer.

Figure 6 shows the equivalent circuit of an LLC converter. To design LLC, the converter is described as follows [3–9].

$$n = \frac{N_p}{N_s} \tag{3}$$

$$m = \frac{L_r + L_m}{L_r} \tag{4}$$

$$f_r = \frac{1}{2\pi \sqrt{L_r C_r}} \tag{5}$$

$$F_x = \frac{f_{sw}}{f_r} \tag{6}$$

$$R_{ac,\,min} = \frac{8}{\pi^2} \cdot \frac{N_p^2}{N_s^2} \cdot R_o \tag{7}$$

$$Q_{max} = \frac{\sqrt{L_r / C_r}}{R_{ac,min}} \tag{8}$$

where N_p and N_s means the turn ratio of the transformer and R_o is the load resistance. Furthermore, f_{sw} is the switching frequency, f_r is the resonant frequency, and the R_{ac} is the reflected load resistance. Therefore, the voltage gain of the resonant tank is defined below.

$$K = \left| \frac{V_{o\_ac}(s)}{V_{in_{ac}}(s)} \right| = \frac{F_x^2(m-1)}{\sqrt{\left(m \cdot F_x^2 - 1\right)^2 + Fx^2\left(F_x^2 - 1\right)^2 (m-1)^2 Q^2}} \tag{9}$$

where $V_{o\_ac}(s)$, $V_{in\_ac}(s)$, Q, m, and F_x are the v_{in_ac}, v_{o_ac}, the quality factor, the ratio of the total primary inductance to resonant inductance, and the normalized switching frequency, respectively. Figure 7 shows voltage gain according to different m and Q factors. In the figure, as the m factor decreases, the voltage conversion ratio is extended. However, according to Equation (4), a low m factor requires low magnetizing inductance L_m. Thus, magnetizing current increases. It means that the conduction loss of magnetic components is increased. In terms of the Q factor, the Q factor is related to the load. The high

Q value means heavy load conditions and the low Q value means light load conditions. Therefore, it is important to set the Q value at the rated load. Thus, a voltage gain of an LLC converter is related to m and Q factors.

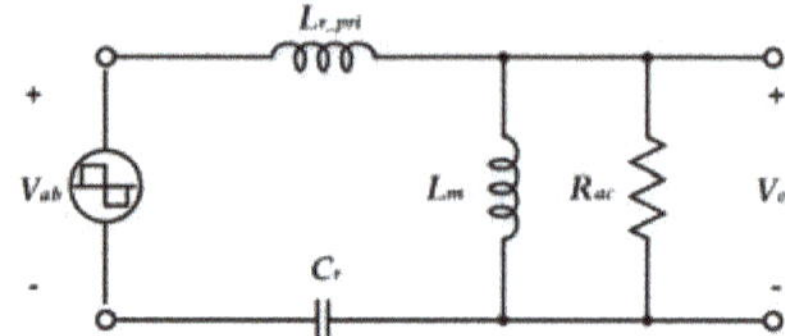

Figure 6. Equivalent circuit of the LLC resonant tank.

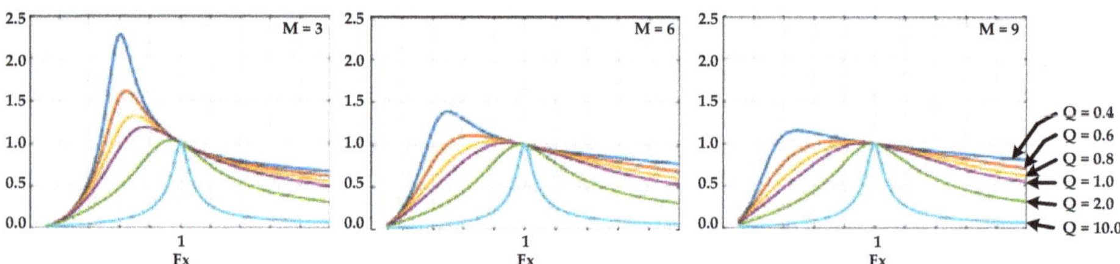

Figure 7. Voltage gain according to different m and Q factors.

4. Simulations and Simulation Results

In order to verify the series-connected devices-based LLC converter, simulation was conducted by Power Sim (PSIM) and Linear Technology spice (LTspice).

4.1. Voltage Balancing Simulation

Figure 8 shows the simulation circuit of voltage balancing. Table 1 shows the parasitic value of components of experiment setup and Table 2 shows the parasitic value of device itself. Voltage balancing simulation was performed by LTspice. Parasitic values of devices were added arbitrarily, and voltage balancing simulation was performed depending on whether the snubber circuit was used.

Table 1. Parameters of the voltage balance test circuit.

Parameters	Values
Input voltage (V_{IN})	1000 V
Gate driver parasitic inductance (L_{g1})	1 pH
Gate driver parasitic inductance (L_{g2})	1.2 pH
Parasitic output capacitance (C_{oss_1})	170 pF
Parasitic output capacitance (C_{oss_2})	100 pF
Inductor (L)	3 mH
Voltage balancing resistor (R_{bal})	1 MΩ
Damping resistor (R_{snub})	5 Ω
Voltage balancing circuits capacitor (C_{snub})	1 nF

Table 2. Parameter of the selected device (C2M0040120D).

Parameters	Minimum	Type	Maximum
Leakage current	×	1 μA	100 μA
Output capacitance		150 pF	

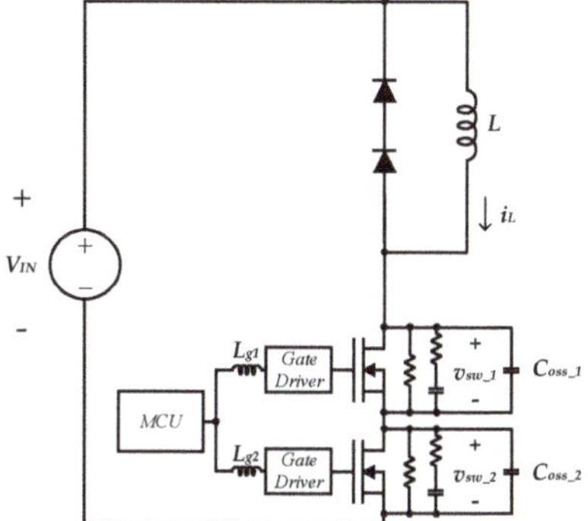

Figure 8. Multi-pulse test circuit for voltage balancing simulation.

Figure 9 shows the simulation results with and without balancing circuits. Without balancing circuits, simulation results show that the voltage balancing between series devices did not converge, and the voltage imbalance increased as the inductor current increased. On the other hand, with balancing circuits, voltage imbalance is somewhat reduced, and the imbalance range is maintained even with the increased inductor current.

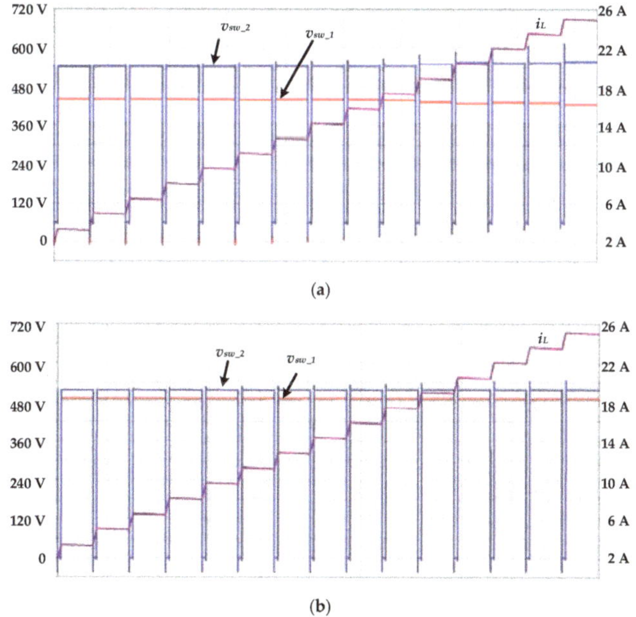

Figure 9. Multi-pulse test simulation result: (**a**) without balancing circuits and (**b**) with balancing circuits.

4.2. LLC Converter Simulation

The LLC converter simulation was conducted on the assumption that voltage imbalance did not occur. The simulation model is shown in Figure 10. Table 3 gives the parameters of the system. An isolated HF transformer can be described as an ideal transformer, magnetizing inductance, and leakage inductance in simulation. In Figure 10, the model of the HF transformer of the LLC converter consists of magnetizing leakage inductance and an ideal transformer, which has a 10:4 transfer ratio. L_m means the magnetizing inductance of the fabricated HF transformer. L_{r_sec} means the secondary side leakage inductance of the HF transformer. L_{r_pri} means the primary side leakage inductance of the HF transformer, which is the sum of the leakage inductance of the HF transformer itself and the added inductance on the outside. In addition, C_r means the resonant capacitor. Resonant capacitors work as a resonant tank with L_{r_pri}. Therefore, the resonant frequency of the LLC converter is determined by the resonant capacitor and the primary leakage inductance of the transformer. Figure 11 shows the voltage gain curve according to the load of the proposed converter.

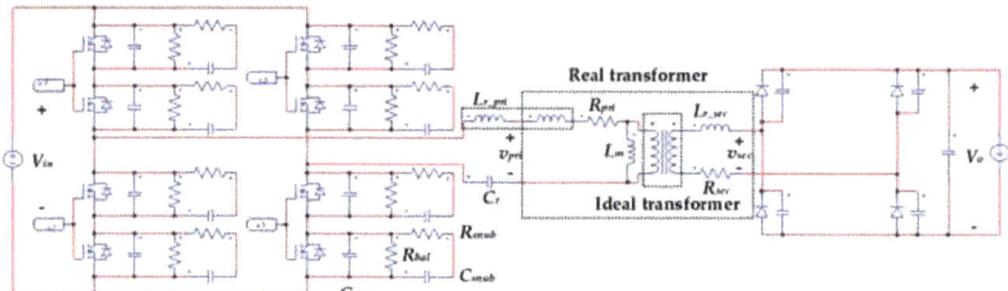

Figure 10. Simulation circuit of series-connected devices based on the LLC converter with a voltage balancing circuit.

Table 3. Parameters of simulation.

Parameters	Values
Input voltage (V_i)	1000 V
Output voltage (V_o)	400 V
Power	3000 W
Resonant frequency (f_r)	100 kHz
Switching frequency (f_{sw})	100 kHz
HF transformer turn ratio (primary: secondary)	10:4
Primary leakage inductance (L_{r_pri})	129 µH
Primary resonant capacitor (C_r)	20 nF
Magnetizing inductance (L_m)	302 µH
Secondary leakage inductance (L_{r_sec})	2.598 µH
Primary resistance (R_{pri})	0.135 Ω
Secondary resistance (R_{sec})	0.110 Ω
Deadtime	500 ns

Figures 12 and 13 show the simulation results of the LLC converter. Figure 12 shows the switching voltage and channel current of the MOSFETs. In the figure, v_{ds_Q1}, v_{ds_Q2}, v_{ds_Q3}, and v_{ds_Q4} represents the drain to source voltages and i_{Q1}, i_{Q2}, i_{Q3}, and i_{Q4} represents the current of devices. In the figure, the channel currents increase after the devices are fully turned on. It means ZVS is achieved. Figure 13 shows the input/output voltage and the current of the HF transformer.

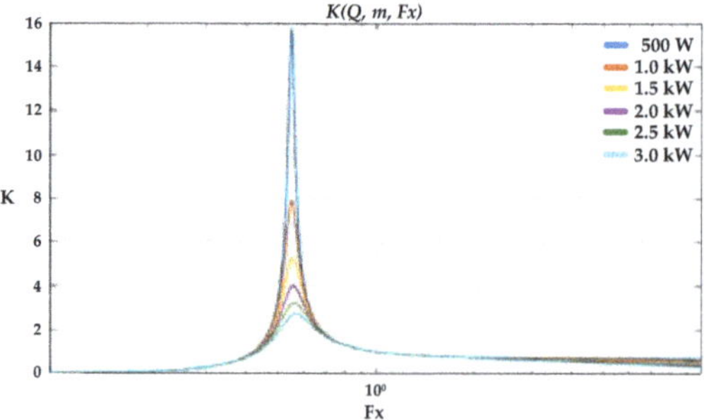

Figure 11. Voltage gain graph of the LLC converter according to a different load.

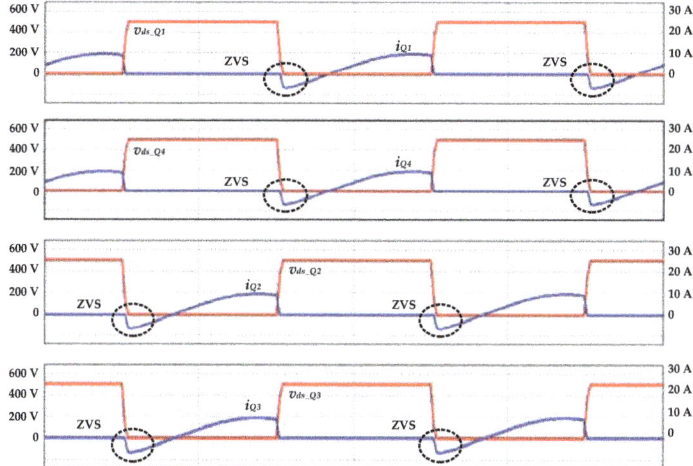

Figure 12. Switching waveform of the drain-source voltage and MOSFET current of the LLC converter.

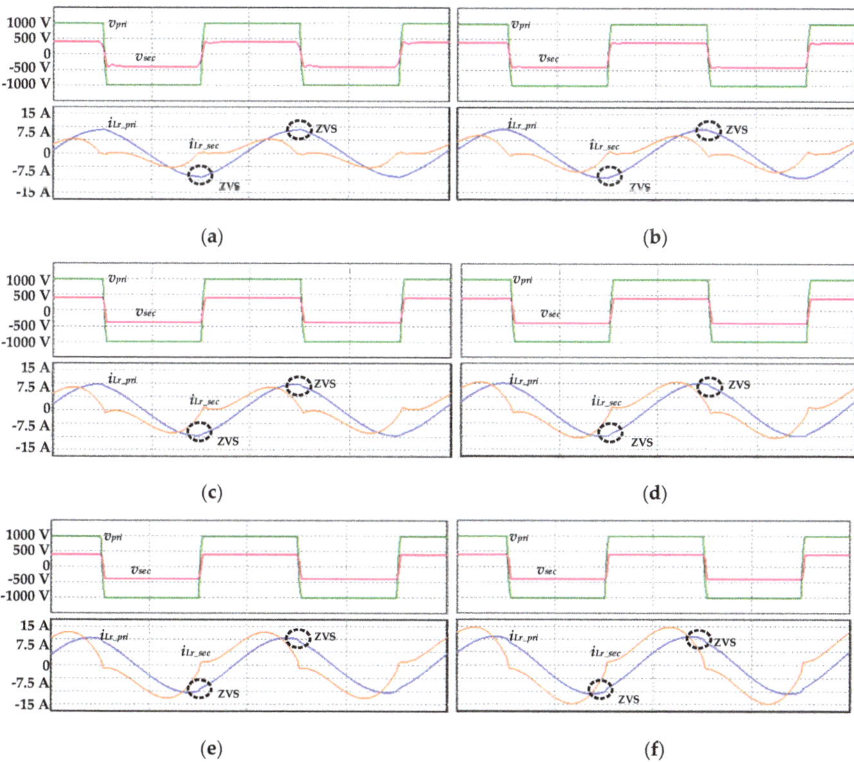

Figure 13. Simulation results of the input voltage v_{pri} and output voltage v_{sec} of the transformer and resonant current of the primary side i_{Lr_pri} and secondary side i_{Lr_sec}: (**a**) simulation at 500 W, (**b**) simulation at 1000 W, (**c**) simulation at 1500 W, (**d**) simulation at 2000 W, (**e**) simulation at 2500 W, and (**f**) simulation at 3000 W.

5. Experiments and Experimental Results

The proposed circuit verified by simulation was tested through experiments. Figure 14 shows the experimental configuration of the proposed LLC converter using a 1200-V SiC MOSFET (C2M0040120D, Cree) and 1200-V SiC Schottky Diode (C4D20120D, Cree). Experimental specifications are shown in Table 3. Before the verification of series-connected devices based on an LLC converter, a multi-pulse test is conducted to verify voltage balancing between series-connected devices.

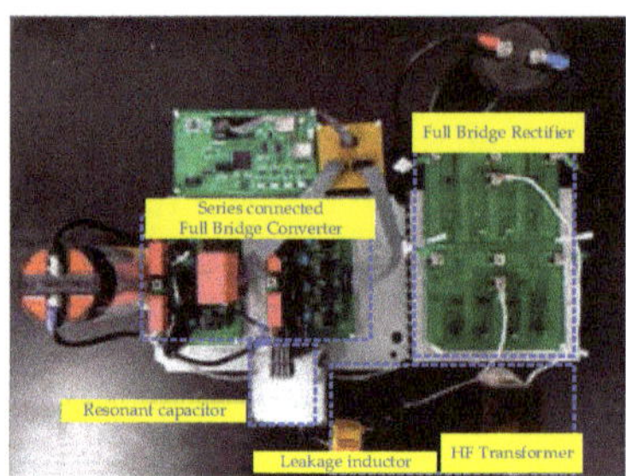

Figure 14. Hardware configuration for a series-connected LLC converter.

5.1. Voltage Balancing Experiment

Voltage imbalance was tested through a multi-pulse test. The experimental setup is shown in Figure 8. The experiment parameters are represented in Table 1.

Figure 15 shows the experimental results without balancing circuits and with balancing circuits. Compared with the simulation results, the trend tends to be different. In the simulation, the factor of voltage imbalance reflects the unbalance of the device's parasitic and the delay of the gate signal. However, according to Reference [12], the parasitic capacitor from gate to ground one of the important factors contribute to a voltage imbalance. In addition, this factor was not reflected in the simulation. Although practical experiments, with the parasitic capacitors considered from gate to ground, the voltage imbalance converges to a steady state through snubber circuits selected by simulation. It means that, if the unbalance voltage does not exceed the rated voltage of the switch, the switch can be regarded as a stable operation with selected balancing circuits by simulation.

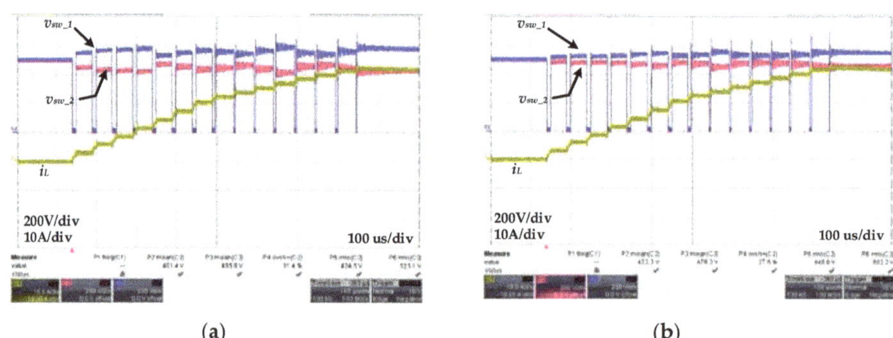

(a)   (b)

Figure 15. Experimental results of the voltage imbalance test: (**a**) without balancing circuits and (**b**) with balancing circuits.

5.2. Double Pulse Test

As mentioned in the previous section, an added snubber circuit for voltage balancing slows down switching speed and increases switching loss. In this section, a Double Pulse Test (DPT) was conducted to analyze switching losses with and without snubber circuits. The experimental setup is shown in Figure 16 and parameters are shown in Table 4. Balancing snubber parameters are shown in Table 1.

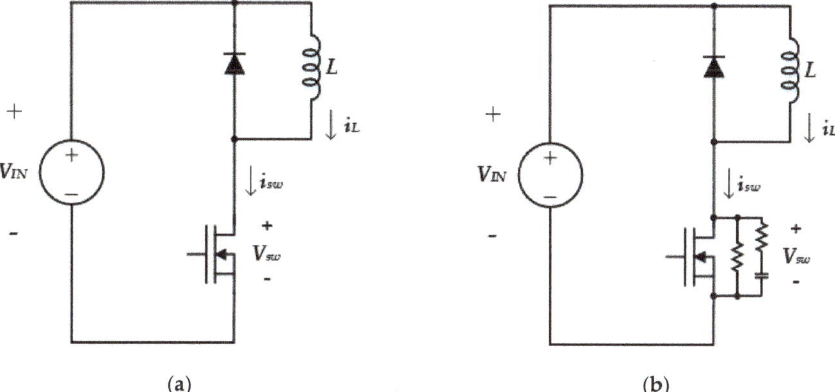

Figure 16. Double pulse test circuits: (**a**) without balancing circuits and (**b**) with balancing circuits.

Table 4. Parameters of the double pulse test.

Parameters	Values
Input voltage (V_{IN})	500 V
Inductor (L)	1.5 mH

DPT was conducted when the DC-link is 500 V. Figure 17 shows DPT results. According to Figure 17, both turn-on and turn-off losses increased with snubber circuits. However, the proposed LLC converter operates ZVS turn-on at all loads. The turn-on loss can be negligible.

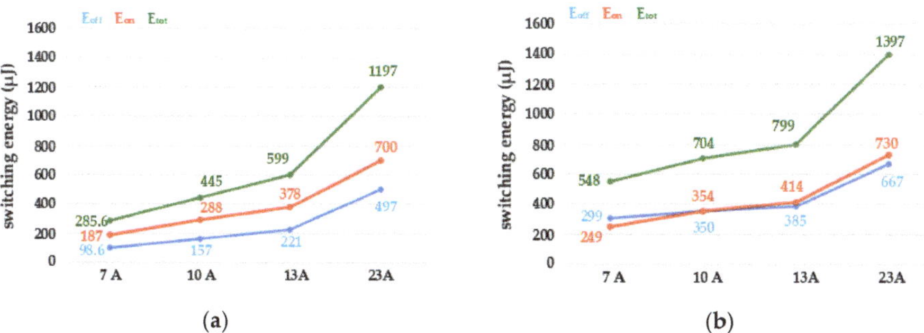

Figure 17. Switching loss comparison according to the switch current: (**a**) without balancing circuits and (**b**) with balancing circuits.

5.3. LLC Converter Experiment

Based on the balancing circuit's value selected through simulations and a multi-pulse test, the series-connected devices based on the LLC converter experiment was conducted. Figure 18 shows the main waveforms of the proposed LLC converter according to the load. In the figure, v_{pri} and v_{sec} represent input/output voltage of the HF transformer, i_{Lr_pri} and i_{Lr_sec} represent input/output current of the HF transformer, according to the load. As shown in the previous section, experimental results were well matched with simulation results.

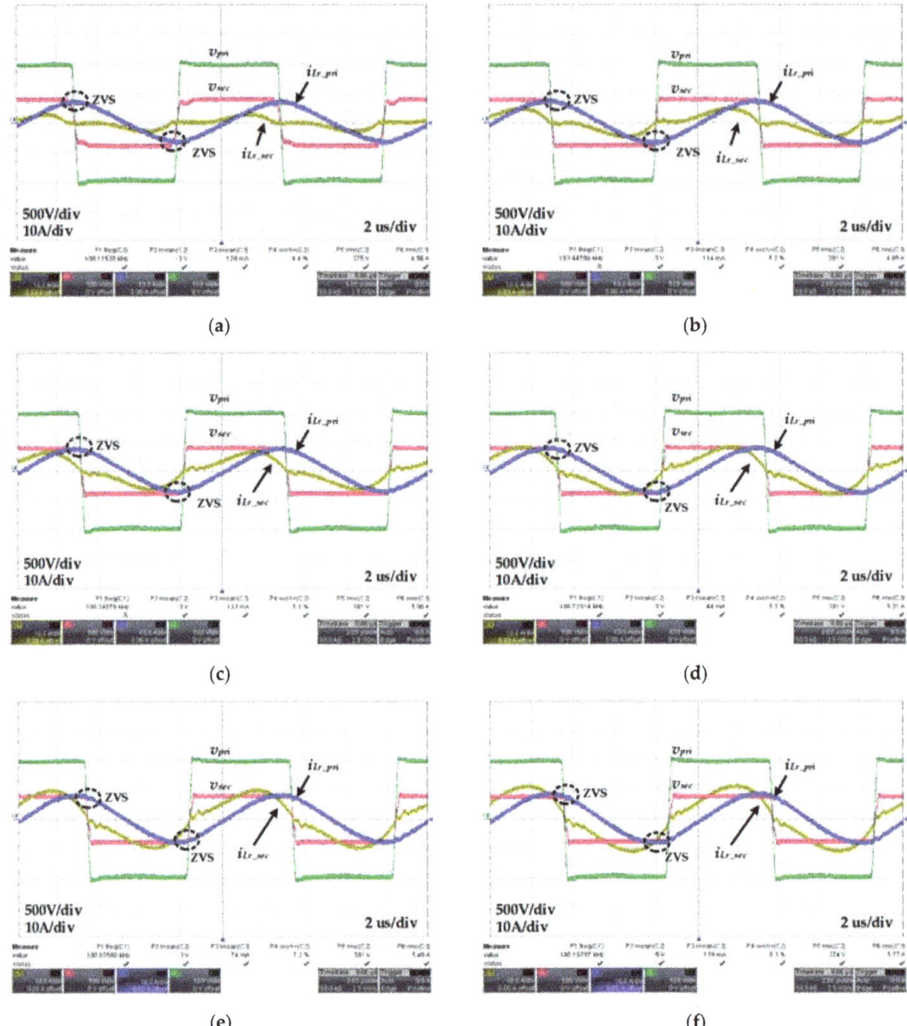

Figure 18. Experimental results of the input voltage v_{pri} and output voltage v_{sec} of transformer and resonant current of the primary side i_{Lr_pri} and the secondary side i_{Lr_sec}. (**a**) Simulation at 500 W, (**b**) simulation at 1000 W, (**c**) simulation at 1500 W, (**d**) simulation at 2000 W, (**e**) simulation at 2500 W, and (**f**) simulation at 3000 W.

Figure 19 shows v_{qs_1}, v_{qs_2}, and v_{gs} at the rated power. v_{qs_1}, v_{qs_2} represent the drain to source voltage of each device connected in series at the rated power. Voltage balancing between series-connected devices did not diverge, and both switches connected in series had achieved ZVS operation. Figure 20 shows the efficiency of series-connected devices-based LLC converter. The maximum efficiency was measured at the rated load.

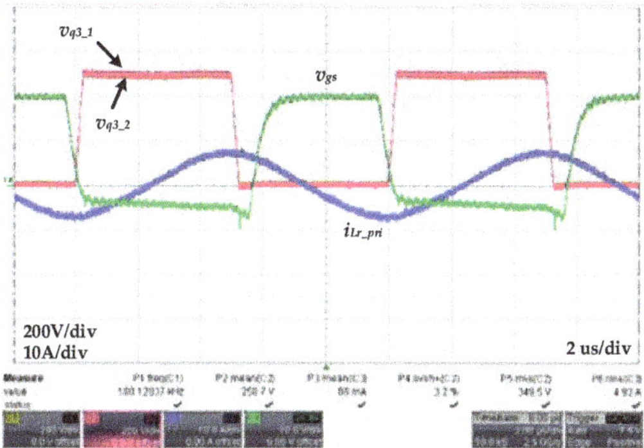

Figure 19. Experimental results of series-connected devices voltage v_{q3_1}, v_{q3_2}, and v_{gs} at the rated load.

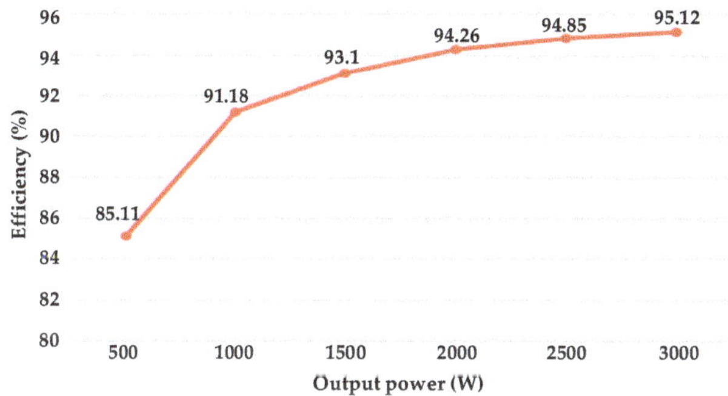

Figure 20. Efficiency of series-connected devices-based LLC converter.

6. Conclusions

This paper describes the design consideration of the fixed frequency LLC converter considered balancing circuit of series-connected devices circuit. In the introduction, the reason for the imbalance voltage across the series-connected devices was described and discussed the need of an LLC converter. The proposed converter consists of a series-connected devices-based LLC converter.

In order to verify the proposed converter, a series-connected devices-based voltage balancing experiment and the series-connected devices-based LLC converter was performed. The snubber circuit, which is the simplest method among the various methods for voltage balancing, was applied and the switching loss comparison and voltage balancing test was performed according to snubber circuits. Furthermore, based on the snubber circuits-designed simulation and experiments, 3 kW series-connected devices-based LLC converter was fabricated and verified. Through the experiments, the voltage balancing between series-connected devices and soft switching operation of the proposed converter are verified. The efficiency of a fabricated converter was measured 95.12% at 3 kW load. When the series-connected based LLC converter is fabricated with balancing snubber circuits, it is confirmed that the LLC converter design is necessary considering the balancing snubber, and the experimental results confirm that the consideration is valid.

Author Contributions: D.Y. implemented the system and conducted the experiments. Y.C. managed the paper. S.L. assisted the idea simulation and the paper writing. All authors have read and agreed to the published version of the manuscript.

Funding: The Human Resources Program in Energy Technology of the Korea Institute of Energy Technology Evaluation and Planning (KETEP) supported this work, which granted a financial resource from the Ministry of Trade, Industry & Energy, Republic of Korea. (NO. 20194030202370) and the Human Resources Program in Energy

Technology of the Korea Institute of Energy Technology Evaluation and Planning (KETEP) granted financial resource from the Ministry of Trade, Industry & Energy, Republic of Korea. (Grant No. 20174010201540).

Conflicts of Interest: The authors declare no conflict of interest.

References

1. Falcones, S.; Ayyanar, R.; Mao, X. A DC–DC Multiport-Converter-Based Solid-State Transformer Integrating Distributed Generation and Storage. *IEEE Trans. Power Electron.* **2013**, *28*, 2192–2203. [CrossRef]
2. Fan, H.; Li, H. High-Frequency Transformer Isolated Bidirectional DC–DC Converter Modules with High Efficiency Over Wide Load Range for 20 kVA Solid-State Transformer. *IEEE Trans. Power Electron.* **2011**, *26*, 3599–3608. [CrossRef]
3. Huang, A.Q. Medium-Voltage Solid-State Transformer: Technology for a Smarter and Resilient Grid. *IEEE Ind. Electron. Mag.* **2016**, *10*, 29–42. [CrossRef]
4. Madhusoodhanan, S.; Tripathi, A.; Patel, D.; Mainali, K.; Kadavelugu, A.; Hazra, S.; Bhattacharya, S.; Hatua, K. Solid-State Transformer and MV Grid Tie Applications Enabled by 15 kV SiC IGBTs and 10 kV SiC MOSFETs Based Multilevel Converters. *IEEE Trans. Ind. Appl.* **2015**, *51*, 3343–3360. [CrossRef]
5. She, X.; Huang, A.Q.; Wang, G. 3-D Space Modulation with Voltage Balancing Capability for a Cascaded Seven-Level Converter in a Solid-State Transformer. *IEEE Trans. Power Electron.* **2011**, *26*, 3778–3789. [CrossRef]
6. She, X.; Yu, X.; Wang, F.; Huang, A.Q. Design and Demonstration of a 3.6-kV–120-V/10-kVA Solid-State Transformer for Smart Grid Application. *IEEE Trans. Power Electron.* **2014**, *29*, 3982–3996. [CrossRef]
7. Shi, J.; Gou, W.; Yuan, H.; Zhao, T.; Huang, A.Q. Research on voltage and power balance control for cascaded modular solid-state transformer. *IEEE Trans. Power Electron.* **2011**, *26*, 1154–1166. [CrossRef]
8. Zhao, T.; Wang, G.; Bhattacharya, S.; Huang, A.Q. Voltage and Power Balance Control for a Cascaded H-Bridge Converter-Based Solid-State Transformer. *IEEE Trans. Power Electron.* **2013**, *28*, 1523–1532. [CrossRef]
9. Hou, K.; Zheng, Y.; Wang, X.; W, L.I.; He, A.; Jiang, Y. Practical Research on VSC Prototype Based on IGBTs Serial Connected Technology. *IEEE Trans. Power Electron.* **2019**, *34*, 495–502. [CrossRef]
10. Ji, S.; Wang, F.; Tolbert, L.M.; Lu, T.; Zhao, Z.; Yu, H. An FPGA-Based Voltage Balancing Control for Multi-HV-IGBTs in Series Connection. *IEEE Trans. Ind. Appl.* **2018**, *54*, 4640–4649. [CrossRef]
11. Ju Won, B.; Dong-Wook, Y.; Heung-Geun, K. High-voltage switch using series-connected IGBTs with simple auxiliary circuit. *IEEE Trans. Ind. Appl.* **2001**, *37*, 1832–1839. [CrossRef]
12. Marzoughi, A.; Burgos, R.; Boroyevich, D. Active Gate-Driver with dv/dt Controller for Dynamic Voltage Balancing in Series-Connected SiC MOSFETs. *IEEE Trans. Ind. Electron.* **2019**, *66*, 2488–2498. [CrossRef]
13. Raciti, A.; Belverde, G.; Galluzzo, A.; Greco, G.; Melito, M.; Musumeci, S. Control of the switching transients of IGBT series strings by high-performance drive units. *IEEE Trans. Ind. Electron.* **2001**, *48*, 482–490. [CrossRef]
14. Ren, Y.; Yang, X.; Zhang, F.; Wang, K.; Chen, W.; Wang, L.; Pei, Y. A Compact Gate Control and Voltage-Balancing Circuit for Series-Connected SiC MOSFETs and Its Application in a DC Breaker. *IEEE Trans. Ind. Electron.* **2017**, *64*, 8299–8309. [CrossRef]
15. Sasagawa, K.; Abe, Y.; Matsuse, K. Voltage-balancing method for IGBTs connected in series. *IEEE Trans. Ind. Appl.* **2004**, *40*, 1025–1030. [CrossRef]
16. Zhang, F.; Yang, X.; Ren, Y.; Feng, L.; Chen, W.; Pei, Y. A Hybrid Active Gate Drive for Switching Loss Reduction and Voltage Balancing of Series-Connected IGBTs. *IEEE Trans. Power Electron.* **2017**, *32*, 7469–7481. [CrossRef]
17. Rahman, S. *Resonant LLC Converter: Operation and Design*; Infineon Technologies North America (IFNA) Core: Durham, NC, USA, 2012.
18. Hu, Z.; Wang, L.; Qiu, Y.; Liu, Y.; Sen, P.C. An Accurate Design Algorithm for LLC Resonant Converters—Part II. *IEEE Trans. Power Electron.* **2016**, *31*, 5448–5460. [CrossRef]

MDPI

St. Alban-Anlage 66

4052 Basel

Switzerland

Tel. +41 61 683 77 34

Fax +41 61 302 89 18

www.mdpi.com

Energies Editorial Office

E-mail: energies@mdpi.com

www.mdpi.com/journal/energies